Managing
Extreme
Technological
Risk

Managing Extreme Technological Risk

Editor

Catherine Rhodes

Centre for the Study of Existential Risk (CSER), UK

World Scientific

NEW JERSEY · LONDON · SINGAPORE · BEIJING · SHANGHAI · HONG KONG · TAIPEI · CHENNAI · TOKYO

Published by

World Scientific Publishing Europe Ltd.

57 Shelton Street, Covent Garden, London WC2H 9HE

Head office: 5 Toh Tuck Link, Singapore 596224

USA office: 27 Warren Street, Suite 401-402, Hackensack, NJ 07601

Library of Congress Cataloging-in-Publication Data
Names: Rhodes, Catherine (Catherine Anne), editor.
Title: Managing extreme technological risk / editor, Catherine Rhodes,
    Centre for the Study of Existential Risk (CSER), UK.
Description: [Hackensack] New Jersey : World Scientific, 2024. |
    Includes bibliographical references and index.
Identifiers: LCCN 2023039688 | ISBN 9781800614819 (hardcover) |
    ISBN 9781800614826 (ebook) | ISBN 9781800614833 (ebook other)
Subjects: LCSH: Technology--Risk assessment. | Risk management. |
    Risk--Sociological aspects. | Science and state.
Classification: LCC TA169.55.R57 M36 2024 | DDC 658--dc23/eng/20240227
LC record available at https://lccn.loc.gov/2023039688

British Library Cataloguing-in-Publication Data
A catalogue record for this book is available from the British Library.

For any available supplementary material, please visit
https://www.worldscientific.com/worldscibooks/10.1142/Q0438#t=suppl

Desk Editors: Nimal Koliyat/Rosie Williamson/Shi Ying Koe

Typeset by Stallion Press
Email: enquiries@stallionpress.com

About the Editor

Catherine Rhodes was Executive Director of the Centre for the Study of Existential Risk (CSER) from 2019–2021, and Academic Project Manager for the Managing Extreme Technological Risk programme from 2016–2020. She retains research affiliations with CSER and the Biosecurity Research Initiative at St Catharine's College, Cambridge. In the context of addressing global risk, her work broadly focuses on understanding the intersection and combination of risk stemming from technologies and risk stemming from governance (or lack of it). Catherine has particular expertise in international governance of biotechnology, biosecurity, and broader biological risk management. She has a background in international relations, but has engaged in extensive interdisciplinary work. Her PhD was funded as part of the "Project to Strengthen the Biological Weapons Convention" at the Bradford Disarmament Research Centre, and Catherine retains a strong interest in international actions to prevent misuse of bioscience.

https://doi.org/10.1142/9781800614826_fmatter

About the Contributors

S. J. Beard is a Senior Research Associate and Academic Programme Manager at the Centre for the Study of Existential Risk (CSER). They work across many areas of existential and global catastrophic risk, including thinking about the ethics of human extinction; developing methods to study extreme, low-probability, and unprecedented events; understanding and addressing the constraints that prevent decision-makers from taking action to keep us safe; and building existential hope in the possibility of safe, joyous, and inclusive futures for human beings on planet earth. Before taking up their current role, SJ worked on Evaluating Extreme Technological Risk as part of the Managing Extreme Technological Risk programme and on Population Ethics in Theory and Practice at the Future of Humanity Institute. They have a PhD in Philosophy from the London School of Economics and serve as a board member for *TANGO Future* and the journal *Futures*.

Haydn Belfield is a Research Associate and Academic Project Manager at the Centre for the Study of Existential Risk (University of Cambridge). Over the past five years, the Centre tripled in size, and he advised the UK, US, and Singaporean governments; the EU, UN, and OECD; and leading technology companies. Haydn is also an Associate Fellow at the Leverhulme Centre for the Future of Intelligence (University of Cambridge). Previously, he worked in UK politics as the Senior Parliamentary Researcher to a Labour MP in the Shadow Cabinet, and was seconded to several general election and referendum campaigns. He was a Policy Associate at the Global Priorities Project (University of

Oxford) and the first Development Director of the Centre for Effective Altruism. He is a DPhil/PhD Candidate in International Relations, and has an MSc in Politics Research and a BA in Politics, Philosophy, and Economics (PPE), all from the University of Oxford.

Adrian Currie is a Philosopher of Science at the University of Exeter, having previously spent two years at the Centre for the Study of Existential Risk as well as having worked at the Universities of Wellington, Calgary, and Sydney, and the Australian National University. Adrian is primarily interested in how scientists successfully generate knowledge in tricky circumstances: where evidence is thin on the ground, targets are highly complex and obstinate, and our knowledge is limited. This has led him to examine the historical sciences — geology, palaeontology, and archaeology — and to argue that the messy, opportunistic ('methodologically omnivorous'), and disunified nature of these sciences often underwrites their success. His interest in knowledge production has also led him to think about the natures of, and relationships between, scientific tools such as experiments, models, and observations, as well as in comparative methods in biology. He also has an interest in how we organise scientific communities, particularly regarding scientific creativity.

Patrick Kaczmarek is the In-House Philosopher at Effective Giving, where he advises ultra-high-net-worth individuals on how they can do the most good. He holds a PhD in Philosophy from the University of Glasgow and is a Research Affiliate at the Centre for the Study of Existential Risk at Cambridge. In the past, he was a Visiting Researcher at the Future of Humanity Institute at Oxford, a Visiting Scholar at the University of Pittsburgh, and taught at the University of Stirling.

Luke Kemp is a Research Associate with Darwin College and the Centre for the Study of Existential Risk at the University of Cambridge. His research focuses on societal collapse and climate change. He has advised the Australian Parliament on ratifying the 2015 Paris Agreement on climate change and has a decade of experience in international negotiations. His work has been covered by media outlets such as *The New York Times* and *The New Yorker*. He is a regular contributor to the *BBC*. He holds a PhD from the Australian National University. Luke is currently writing a book on societal collapse and transformation (titled *Goliath's Curse*) which will be published with Penguin (Viking Books) in 2024.

Matthijs M. Maas is a Senior Research Fellow (Law & AI) at the Legal Priorities Project and a Research Affiliate at the Centre for the Study of Existential Risk, University of Cambridge. His work focuses on mapping theories of change for transformative AI governance, different international institutional designs for AI, the effect of AI on international law, and arms control regimes for military uses of AI technology. Matthijs received a PhD in Law from the University of Copenhagen and an MSc in International Relations from the University of Edinburgh. He has previous experience working at The Hague Centre for Strategic Studies and the Dutch Embassy in Beirut, amongst others.

Huw Price is Emeritus Bertrand Russell Professor of Philosophy and an Emeritus Fellow of Trinity College, Cambridge. He was Co-Founder with Martin Rees and Jaan Tallinn of the Cambridge the Centre for the Study of Existential Risk and Academic Director (2016–2021) of the Leverhulme Centre for the Future of Intelligence. Since 2022, he has been a Distinguished Professor Emeritus at the University of Bonn, Germany.

Lalitha S. Sundaram is Senior Research Fellow at the Centre for the Study of Existential Risk (CSER), University of Cambridge, UK. Her work at the CSER is in the area of biorisk, with a particular emphasis on regulation and governance. She investigates risks — real or perceived — surrounding emerging biotechnologies such as synthetic biology. Before joining CSER, Lalitha worked at the University of Cambridge and Edinburgh's Arsenic Biosensor Collaboration, where she developed a strategy to take this novel synthetic biology product from bench to field, focusing on the international regulatory landscape and Responsible Research and Innovation. Following this, she held a fellowship at King's College London investigating the opportunities and challenges facing emerging biotechnologies seeking to tackle global health challenges. Lalitha's PhD research, also at the University of Cambridge, used a combination of bioinformatic, next-generation sequencing, and molecular biology tools to explore host–cell metabolic and microRNA changes following infection by the pathogenic parasite *Toxoplasma gondii*.

https://doi.org/10.1142/9781800614826_fmatter

Contents

List of Figures

List of Tables

List of Abbreviations and Acronyms

AAAI	Association for the Advancement of Artificial Intelligence
ABM	Anti-Ballistic Missile
AGI	Artificial General Intelligence
AI	Artificial Intelligence
AI HLEG	High-Level Expert Group on Artificial Intelligence to the European Commission
ARPA-E	Advanced Research Projects Agency-Energy
ASOGR	A Science of Global Risk
BEC	Brillouin Energy
BWC	Biological Weapons Convention
CAHAI	Council of Europe's Ad Hoc Committee on Artificial Intelligence
CBP	Customs and Border Protection
CCW	Convention on Certain Conventional Weapons
CDEI	Centre for Data Ethics and Innovation
CEN-CENELEC	European Committee for Standardisation and the European Committee for Electrotechnical Standardisation
CMNS	Condensed Matter Nuclear Science
COP	Coefficient of Performance
COP26	26th Conference of the Parties to the United Nations Framework Convention on Climate Change
CSER	Centre for the Study of Existential Risk

CSET	Centre for Security and Emerging Technology
DARPA	Defense Advanced Research Projects Agency
DIA	(US) Defense Intelligence Agency
DMA	Digital Markets Act
DOE	Department of Energy
DSA	Digital Services Act
DURC	Dual-Use Research of Concern
EA	Effective Altruism
ENEA	Energia Nuclear ed Energie Alternative (Atomic Energy and Alternative Energy) Agency
ERV	Expert Responsible for Validation
ETR	Extreme Technological Risk
FAO	Food and Agriculture Organization
FAT ML	Fairness, Accountability, and Transparency in Machine Learning
FTC	Federal Trade Commission
GCR	Global Catastrophic Risk
GGE	Group of Governmental Experts
GPAI	Global Partnership on Artificial Intelligence
GPT	General-Purpose Technologies
IAP	InterAcademy Panel
IARPA	Intelligence Advanced Research Projects Agency
ICCF	International Conference on Cold Fusion
ICE	Immigration and Customs Enforcement
ICML	International Conference on Machine Learning
IDEA	Investigate, Discuss, Estimate, Aggregate
IEEE	Institute of Electrical and Electronics Engineers
IH	Industrial Heat
IJCAI	International Joint Conferences on Artificial Intelligence
ITN	Importance, Tractability, and Neglectedness
ITU	International Telecommunications Union
KAIST	Korea Advanced Institute of Science and Technology
LAWS	Lethal Autonomous Weapons Systems
LCFI	Leverhulme Centre for the Future of Intelligence
LENR	Low-Energy Nuclear Reactions
METR	Managing Extreme Technological Risk
MILA	Montreal Institute for Learning Algorithms

ML	Machine Learning
MOOC	Massive Open Online Course
MWI	Moral Worthwhileness Intuition
NASA	National Aeronautics and Space Administration
NDA	Non-Disclosure Agreement
NeurIPS	Neural Information Processing Systems
NGO	Non-Governmental Organisation
NIST	National Institute of Standards and Technology
NRC	National Research Council
NSC	National Security Council
OECD	Organisation for Economic Co-operation and Development
OSTP	Office of Science and Technology Policy
RCR	Responsible Conduct of Research
RI	Responsible Innovation
S&T	Scientific and Technological
SALT	Strategic Arms Limitations Talks
SPAWAR	Space and Naval Warfare
TERRA	The Existential Risk Research Assessment
TEVV	Test and Evaluation, Validation and Verification
TUC	Trade Unions Congress
TWCF	Templeton World Charity Foundation
UBC	University of British Columbia
VBM	Valuable Biological Material
WEXD	When Experts Disagree
WHO	World Health Organization
WOAH	World Organisation for Animal Health

Chapter 1

Introduction to Managing Extreme Technological Risk

Catherine Rhodes

Humanity is faced by a class of risks that crosses a threshold above which its entire existence is threatened. These are often referred to as 'existential risks'. A further set of risks, known as 'global catastrophic risks', may take us over this threshold if they occur in combination, cumulatively, or in cascades. Such risks also threaten to substantially harm humanity even if they occur in isolation.

The extreme risks covered in this book encompass existential and global catastrophic risks. While there is a particular focus on the extreme risks with a technological source, as Chapter 2 explains, there are few such risks without a significant technological dimension whether that be through their root cause, their spread mechanisms, or the required responses.

Managing Extreme Technological Risk was the first, and foundational, major research programme of the Centre for the Study of Existential Risk (CSER).

This book does not seek to outline and explore particular extreme risks — there are other works which tackle this dimension well (recent examples include Toby Ord's *The Precipice*, Olle Häggström's *Here Be Dragons*, as well as Martin Rees' *Our Final Century* on which the work of Managing Extreme Technological Risk explicitly builds). Instead, this is about the journey taken by a set of researchers coming together across disciplines to demonstrate the type of programme that can develop our understanding of this class of risks, their management and mitigation.

It outlines a number of challenges we faced as we explored this space, and challenges in terms of epistemology, evaluation, research, and action as we moved towards a 'science' for examining extreme technological risk in a robust way that could provide grounds for and routes to affect change, particularly in the way policy and technology communities approach such risks and participate in their management.

This emphasis on challenges should not been seen as a negative. The fact that we collectively had space and opportunity to acknowledge, examine, consider, and test appropriate responses to these challenges (and in works such as this — reflect on and communicate about them) is likely to have high value for the field of existential and global catastrophic risk research, as similar exercises would for other emerging fields. By making such activities part of our practice, we also align with ideas and frameworks that we believe have utility in the management of extreme technological risk (such as responsible innovation — see Chapter 6), so we also have the value of holding ourselves to the standards we expect from others.

The book includes contributions from our researchers, research associates, and one of one of CSER's founders and the Principal Investigator for the programme, Professor Huw Price. While drawing primarily on experiences of the Managing Extreme Technological Risk research programme, it also reaches beyond that, in some cases highlighting what led to the programme, in others picking up on key questions and challenges that remain. Work in this essential and urgent area is still in its early phases; it will retain relevance and significance for decades to come and into humanity's long-term future. The communities engaging with and contributing to the field continue to grow rapidly and the lessons learned and reflections from this book will not only be of particular interest to them but also contain insights useful to those working in aligned fields (of research, policy, and practice), and to other areas in which efforts to create truly interdisciplinary initiatives addressing major global challenges are occurring.

Chapter 2, Extreme Risk and the Culture of Science: Two Challenges, provides a concise outline of what extreme technological risk is, and a framework for understanding its technological dimensions, including how it permeates and spreads through social and technical systems and networks. It goes on to outline the challenges that arise for a developing field that needs to embrace ambiguity due to the nature of knowledge available to, and able to be produced by, it. It leaves open questions of whether practice in the field can be structured in such a way to enable an emphasis on ambiguity, and whether there are appropriate incentives available for this. As Sundaram and Currie observe, the challenge 'is to develop a set of standards which are well adapted for a science of ambiguity, and particularly for understanding such risks. Incentives should be geared towards understanding, and systematising, spaces of possibility … [to] encourage this kind of open-ended exploration while nonetheless providing common grounds and baselines for research to accumulate.'

Chapter 3, Risk and Scientific Reputation: Lessons from Cold Fusion, dives deeper into the topic of how scientific standards that are ill adapted to low-probability, high-impact scenarios (be that impact in terms of extreme benefit or extreme harm to humanity) may lead us to ignore insightful voices until a point at which effective interventions come too late, or at significantly greater cost to humanity. It explores, through detailed cases, the ways in which the epistemic challenges associated with extreme technological risk and conventional scientific standards present problems for individuals within scientific and technological communities, should they wish to raise concerns about potential extreme risks associated with their field. In this sense, it is not directly about challenges faced by research within the field of existential risk, but about those faced by others whose knowledge and insights may be extremely valuable in identifying and understanding sources of technological risk. However, the challenges are rooted in the same underlying issues identified in Chapter 2 — of scientific standards that are mismatched or ill adapted for dealing with ambiguity. In some ways, the challenge is more severe when it occurs in scientific fields in which such standards are deeply entrenched, whereas as an emerging field, the study of extreme technological risk at least has a window of opportunity to build in standards that support ambiguity and creativity as it develops.

Chapter 4, Foreseeing Extreme Technological Risk brings understanding of how methods and techniques for anticipation can best be adapted to encompass the ambiguity of a well-adapted science of extreme technological risk. It provides clear examples of the challenge described in Chapter 2 — of methods, based on scientific standards that place high value on certainty and/or accuracy, being poorly adapted for developing knowledge about extreme risks. It describes what it has been like in practice to face and try to address this challenge through use of foresight methods and the careful, steady iteration of these towards something more capable of capturing systems, interconnections, and key intervention points.

While some foresight techniques appear to be a good fit for a well-adapted science of extreme technological risk, it is acknowledged that there remains limited evaluative information and analysis about what works well among these approaches. Perhaps this speaks to another layer of ambiguity that a science of extreme technological risk needs to embrace: we may achieve greater evaluative information over time, but were we to wait until we comprehensively understand the effectiveness of the various techniques before deploying them, we would leave ourselves in a similar situation to that outlined at the start of Chapter 4 — blind to risks until a point at which intervention is too late.

Chapter 5, Evaluating Extreme Technological Risk: A Social Contract Based Approach, responds to an epistemic challenge raised by the nature of the field as one which — to be successful — needs to engage across disciplines, population groups, and areas of practice, and thus needs to incorporate an additional form of 'epistemic humility' to that suggested in Chapter 3. The authors of this chapter — who have done a lot of work on evaluation of extreme technological risk from a philosophical perspective — begin to develop an evaluative approach that can 'constructively engage with other moral perspectives', recognising the value of this for a field that necessarily needs to connect with multiple stakeholders, not all of whom are comfortable with consequentialist approaches.

It was integral to the programme to not only bridge and combine disciplines but also connect to policy and technology communities whose actions and activities can substantially influence the nature of the risks we face and the opportunities we have to manage them. The close link between consideration of how to study extreme technological risk and how to connect the understanding and knowledge developed with policy and other social processes is outlined towards the end of Chapter 2, and the theme can be traced through several of the other chapters. Reflections in Chapter 4 lead to questions about the type of 'reputation traps' political leaders may face when presented with perverse incentives, reinforcing points made in Chapter 2 about conceptual influence. For Chapter 3, such influence needs to also extend to scientific incentive systems, by encouraging more open-mindedness and constructive dialogue, avoiding epistemic slurs. In turn, Chapter 6, Responsibility and the Management of Extreme Technological Risk: Bio(techno)logical Risk, suggests that such open-mindedness, including abilities to self-reflect, anticipate, and be flexible, is key to notions of responsibility in managing extreme technological risk, and Chapter 4 notes that part of the focus of foresight activities is the cultivation of anticipatory and open mindsets among participants. While Chapters 6 and 7, A Decade of Responsible Innovation by the AI Community 2012–2022: Analysing Recent Achievements and Future Prospects, have a more outward-looking focus in their discussions of responsibility, it is clear throughout the book that the key aspects of responsible science apply as much to the work we do ourselves as to other communities we engage with, and that they should continue to guide us as the field develops further, particularly in ensuring that space remains for reflexivity and epistemic humility.

Two key areas of technology risk that Managing Extreme Technological Risk focused on were advanced biotechnologies (reflected in Chapter 6) and AI (discussed in Chapter 7). Some of our researchers spent periods of time 'embedded' with related technology communities. Chapter 7 brings together insights from such engagement with analytical frameworks to explore the role that activism by the AI community can play in relation to responsible innovation. The chapter also assesses the role activism can play in responsible AI development through applied understanding of epistemic communities, worker organising, and veto player framings, while Chapter 6 reflects on connections between the concepts of the web of prevention, responsible science, and a framework of responsible innovation. Together, these chapters demonstrate the usefulness of drawing on concepts and approaches from different disciplines to gain greater understanding of routes to management of extreme technological risk.

Chapter 8, From Evaluation to Action: Ethics, Epistemology, and Extreme Technological Risk, closes the book with a further reflective piece that takes us onwards in our journey as researchers within an emerging field. Drawing on the earlier chapters and the individual experiences of its authors as researchers in this field, they consider further some of the epistemological and ethical challenges they've encountered and explore seven key questions these experiences raise. Their responses to these questions provide a starting point for approaching these challenges. However, in keeping with the desire for epistemic humility and an embrace of ambiguity in the development of a 'science' of managing extreme technological risk, they do not seek to provide definitive answers but to prompt all those working within this area to keep space for reflection and development of constructive responses that can sustain this field long into the future. Rather than simply providing a direct understanding of extreme technological risk and its management, what has emerged throughout the contributed chapters is a profound connection between our encounters and the larger questions surrounding the essence of 'science'; what it is, should be, and can be in terms of meeting humanity's most pressing needs. The challenges we face as researchers within this specific field are likely also to be faced by many researchers outside it, and just as this book provides useful insights relevant beyond the existential risk community, we will continue to benefit from learning from those outside it who face similar issues.

Chapter 2

Extreme Risk and the Culture of Science: Two Challenges

Lalitha S. Sundaram and Adrian Currie

2.1 Introduction

The systematic — scientific — study of extreme technological risk is both urgent and tricky. Urgent because as our technological prowess increases, so too do the potential impacts (intended or otherwise). Understanding the scope of those risks, how to mitigate and avoid them, is critical for powerful, transformative technologies leading us to brighter rather than darker (or indeed snuffed-out) futures. It is tricky, we think, due to the nature of the knowledge we might have of those risks.

Extreme technological risk, like other catastrophic or existential risks, combines low probability with high impact (Bostrom, 2013; Millett and Snyder-Beattie, 2017; Ó hÉigeartaigh, 2017). High-impact events interrelate with many complex, global systems in bewildering ways, making modelling and understanding the relevant causal chains incredibly difficult. Generating evidence pertinent to understanding low-probability events is challenging due to their rarity, the complex systems they affect, and the long temporal scales at hand. Further, the technological aspects of such risk make its occurrence more dependent on human decisions and social factors — themselves difficult to predict — than other kinds of low-probability, high-impact risks.

These elements make a difference to the kind of outputs we should expect from a science of extreme technological risk. We'll argue that instead of precise predictions or even attributions of probability, such a science should fundamentally embrace *ambiguity*. In particular, it should be in the business of tracking the space of possible futures related to extreme technological risk, and ascertaining the kinds of causal factors which could lead to, from, or mitigate the impacts of, such risks. We'll highlight two challenges arising for an ambiguous science of extreme technological risk.

First, a science of extreme technological risk must develop *standards* that are geared towards appropriate epistemic outcomes. Where many scientific standards are geared towards stating what is certain (by having sufficient statistical significance, for instance), a science of extreme technological risk requires standards geared towards ambiguity: laying out and exploring spaces of possibility. Second, researchers of extreme technological risk (those in AI ethics, existential risk, etc.) are typically not only interested in understanding such risks for their own sake but also interested in influencing policy; the point of knowing what factors will pull us towards safer futures is to try and get those factors put into place (Beard and Torres, 2020; Sears, 2020). However, the kind of knowledge we expect the science of such risks to produce requires packaging to be salient to policymakers, and there are various ways in which the epistemic status of such a science might undermine this.

Moving forwards, we'll first characterise why a science of extreme technological risk will be a science of ambiguity, before articulating our two challenges. In this short chapter, our aim is geared more towards characterising the science at an abstract level and articulating the challenges, rather than providing a positive picture of how such challenges can be overcome.

2.2 Ambiguous Science

Extreme technological risk is low probability but high impact: the chances of its occurring are — at least in the short term — low, but the consequences of its occurring are extremely dire. In this section, we suggest that the knowledge we might have of such risks is better characterised in terms of *ambiguity* than *certainty*. We take a specific notion of 'ambiguity' from archaeological theory.

The archaeologist Joan Gero has argued that her field goes astray when it emphasises what is certain at the expense of what is ambiguous (Gero, 2007). Archaeologists make inferences into the cultural past of our species via their partial, degraded remnants. As the range of human expression, culture, and lifeways is so vast, and the archaeological record so incomplete, the evidential base that archaeologists work with often provides little constraint on their hypotheses. And yet, as Gero (2007, 312) puts it:

> Even when [archaeological] reports are qualified by degrees of probability and tempered by calls for more data, it is certainty that fills the literature and characterizes how archaeological results are reported in our grant proposals, conference papers, journal articles, and in the popular press ... Why is there so little discussion devoted to the other matter that matters so much: that every phase and feature of archaeological research requires archaeologists to make difficult or even impossible interpretive decisions on the basis of incomplete, unfamiliar, indeterminate or bewilderingly complex evidence? Most often, the confusion, uncertainty and ambiguity are left out of our conclusions, overlooked, ignored, forgotten or erased.

What is the difference between reporting *certainty* and reporting *ambiguity*? A science geared towards certainty tells us which hypothesis is the most likely, or which hypotheses do better than the null, say. Generally, they are interested in establishing what is *actual*; 'given the evidence we have, what can we say for sure?' A science of ambiguity tells us, given the available constraints, which hypotheses are still open and what questions still remain. Generally, they are interested in exploring what is *possible*. Of course, certainty and ambiguity are not necessarily in conflict: knowing what is possible is often relevant for establishing what is actual. But Gero's point concerns what is *rewarded* or *aimed at*: what the structure of research practice targets. Her worry is that archaeological practice is incentivised towards emphasising the certain at the expense of the ambiguous.

Studying archaeology and studying extreme technological risk are highly different pursuits (the former being interested in uncovering past cultures, the latter in exploring future outcomes) but they are unified by a crucial feature: hypotheses that lack strong empirical constraints.

We'll emphasise three factors contributing to this, but begin by distinguishing between existential risks generally and those arising from technology.

Existential risk is sometimes, especially in lay contexts, crudely categorised as either being anthropogenic or 'natural' (the former typically seen as technology-driven), but this distinction quickly falls apart when we consider how thoroughly technology pervades so-called 'natural' risks. Pandemics, for instance, are often considered 'natural' existential threats. However, infections only become global pandemics when pathogens access vast, complex — decidedly 'non-natural' — human networks. Moreover, pandemics only become existential threats when their *impact* propagates through even more complex human networks, themselves often highly technology-dependent.

Although we don't think a naïve 'natural/anthropogenic' dichotomy can distinguish extreme technological risk from other extreme risks, we nonetheless think focusing on the technological aspects of risks highlights particular epistemic challenges. For instance, Avin *et al.* (2018) introduce a particularly useful framework for classifying existential risks:

> (i) a critical system (or systems) whose safety boundaries are breached by a potential threat, (ii) the mechanisms by which this threat might spread globally and affect the majority of the human population, and (iii) the manner in which we might fail to prevent or mitigate both (i) and (ii).

So, an existential risk can be understood in terms of, first, the boundaries that are breached, second, how the threat from such a breach might propagate, and third, how attempts to block or mitigate the initial breach or propagation might fail. These three facets of existential risk are particularly relevant when it comes to extreme technological risk, given that our relationship with and use of technological elements is so intertwined with each. Mishaps or malicious use of powerful technologies is a clear way of breaching critical systems; technology underwrites many ways of propagating system collapse, and the lack of technology or its misuse may lead to failures of mitigating system collapse or propagation. Technology can be the source of the threat itself, but in nebulous and long-term ways. Although this doesn't provide a strict distinction between technological and 'natural' existential risks, it does allow us to highlight our target. Asteroid impacts and supervolcanic eruptions might be mitigated or avoided by technological intervention and system design, but many of the systems breached and mechanisms of propagation are not at base techno-logical. Engineered biogenic threats, as well as those from artificial intelligence, are. In the archetypal view of the relationship of risk with hazards, exposures, and vulnerabilities, extreme *technological* risk there-fore holds a special place in that our use of technology can drive all three components. Anthropogenic climate change and pandemics sit some-where in the middle.

A science of extreme technological risk, then, can be distinguished not only from generalised notions of 'science' but also from studies of other kinds of existential risk. How we study asteroids and solar flares will necessarily differ from how we explore the possible futures of AI or biotechnology. For a start, there is far greater ambiguity about the nature of these latter threats, whether and how we can respond to them. So, given the nature of extreme technological risk, what factors influence our epis-temic situation concerning them?

First, extreme technological risk is often low probability, at least over relatively short periods of time, and this makes directly relevant data rare. A common way of getting an empirical grip on some class of events is to examine other instances or analogues of those events. Say we want to understand the probability of some event *e* occurring given some set of background conditions *c*. A good approach is to survey instances of *e-type* events across instances of *c* and not-*c*. How often when *c* occurs do *e-type* events also occur? This allows us to establish a correlation between the two. By further examining *e-type*s across other ranges of conditions, we can begin to infer from correlation to causation, or at least to grasp a statistical relationship between background and event. However, what if there are no or very few *e-type* events, or no or very few *c*s available? Under such conditions, this strategy becomes extremely weak. This alone should not lead to pessimism about our capacity to come to know things about *e* or *e-type* events — there are multiple routes to knowledge even of putatively unique events (see, for instance, Buskell and Currie, 2021) — but it remains that one powerful part of our arsenal of scientific strategies is unavailable.

Second, extreme technological risk is *high impact*. We understand 'impact' in terms of both the event's magnitude and its ramifications, rather than magnitude alone. Typically, the set of circumstances and events that have the highest impacts will strike at multiple scales affect-ing many systems. Some could be multiple small catastrophes that reverberate through a variety of systems, some could be more like cascades

(Kareiva and Carranza, 2018; Liu *et al.*, 2018; Manheim, 2020). Whether we are thinking of single major events or of 'perfect storms' involving complex cascades, the majority of existential risks — or at least the ones that are anthropogenic or to a degree under human influence — will see their effects felt across networks. To understand the results of a high-impact event, then, we must understand how it interrelates with a wide variety of systems, themselves interrelated. We have seen how the impacts of decidedly non-existential threats manifest dramatically across interconnected systems, such as the eruption of the Icelandic volcano Eyjafjallajökull in 2010 or the grounding of the Evergreen cargo ship in the Suez Canal in 2021. Existential threats, due to their magnitude and wide-spread ramifications, would be even more perniciously embedded across complex networks. And if our ignorance of each system impacts our overall capacity to understand the effects — and causes — of that event, then our epistemic challenge also increases.

Third, extreme technological risk is understood at a long temporal scale. At longer temporal scales, our ignorance of which systems are relevant, how they may have changed, and even which systems will be active increases. Things we can take for granted at a shorter scale come up for grabs as we extend our reach outwards. Thus, our ignorance further compounds. Although any event with a probability higher than zero will be guaranteed to occur as we approach infinite time, this is no panacea to our epistemic situation; we can only examine events within a desperately small temporal window. This is only exacerbated when we consider that our technologies will change these systems dramatically and unpredictably, and so a science of extreme technological risk becomes indeterminably more complex (Manheim, 2020; Pegram and Kreienkamp, 2019).

This compounding of ignorance is sometimes framed as a sociological problem in science governance, but it applies to the study of extreme technological risk as well. Consider, for example, Collingridge's 'dilemma of control' (originally stated in 1980 in his book, *The Social Control of Technology*), since elaborated upon (see, for instance, Genus and Stirling, 2018). Collingridge claims that the earliest points in the development of a technology are those with the least available information regarding the technology's consequences and potential impacts. However, this is exactly the time at which intervention is most effective or even possible at all. As technology becomes entrenched, later interventions become difficult: the impacts are beginning to manifest and the 'course is set'. As we have noted, the long term is what concerns us when it comes to existential risk, while our study (and our influence on technology policy, see the below paragraph) is largely seen to exist in the present and near terms.

The upshot of all this is that a science of extreme technological risk is best understood as, and approached in terms of, a science of ambiguity: instead of targeting the actual, we should target the possible. Given what few constraints we have or know about, what outcomes can be excluded, which remain open, what possible futures are available, and so on, how can we shift towards desirable futures? We can attempt to systematise this research to an extent (by using the Avin *et al.* (2018) framework, for example) to explore what connects our critical systems, what common vulnerabilities they display, how a threat might spread, and how we might prepare and respond. Much of this will need to be done attempting to identify dependencies between various background conditions and outcomes, what Liu *et al.* (2018) term 'existential vulnerabilities' or 'existential exposures'.

Indeed, what we do when we study extreme technological risk is describe a set of possible interconnected futures. Horizon scanning and foresight as described in Chapter 4 are useful methodologies that we can use to formalise this kind of exploratory research. Here, the aim is not to generate a predictive checklist of hypotheses to be falsified, or even probabilities to be precisely quantified. The value in that research is in many ways the very process of doing the exploration and imagining. Such systematic imagining of possibilities serves many functions: it can form the basis of understanding just how various complex events and systems might interrelate, what signals might be warning signs of danger or of safer waters, and how — and when — to act to mitigate extreme technological risk.

So, a science of extreme technological risk ought to be a science of ambiguity. But what would be required for such a science to flourish? In the remainder of this chapter, we'll point to two challenges. The first is related to the internal organisation of science: how we decide what to pursue and how this is incentivised. The second asks how such results might influence policy.

2.3 Standards

There are at least two kinds of questions we might ask after the organisation of a scientific discipline. First, following the philosopher

Philip Kitcher, we might ask whether the science is *well-ordered* (Kitcher, 2003; Cartwright, 2006). This is to ask whether the research directions — what investigations are carried out — are the right ones. And this is determined largely as a function of three elements: the desirability of a particular line of enquiry (research that is 'of public benefit' being a modern way of phrasing this); the costs of pursuing particular research agendas; and the likelihood of their achieving their ends, with these latter two being balanced against each other quite explicitly. (NB 'cost' here refers to any kind of resource and includes ethics). What counts as 'right' involves a trade-off between the probable success of such investigations, their cost, and the desirability of success. For Kitcher (2003), the latter is best determined democratically, in terms of public needs and desires.

Cost–benefit analysis is by no means foreign to the study of extreme technological risk (see Chapter 5 for a more thorough discussion); however, such discussion usually refers to the cost (the risk to humanity) of pursuing a particular line of technological development, balanced against what benefits that same technology might afford. For example, is research that involves dangerous pathogens worth the risk, however remote, of a laboratory leak? In this chapter, as opposed to the risks and benefits of pursuing technological development, we're interested in what should set the research agenda of existential risk as a field, rather than of the dangers and benefits of researching particular technologies.

A well-ordered science of extreme technological risk might therefore be a science where some method of prioritisation points us to the risk that is of greatest benefit to study and then refines this list based on the subset that is 'economically sound' (that is, it balances cost, chance of success, and desirability of success). But for existential risk, this is problematic for a number of reasons: the essential ambiguities that we have described above (low probability, high impact, complexity, and long timeframe) render any attempt at prioritisation (such as seen on 'risk registers' that map likelihood against severity) either too easily 'proven wrong' to garner credibility or too broad to be truly useful (see Beard *et al.*, 2020, for a review of quantified methods). How can we decide which risk is the most worthwhile to study? The one with greatest impact, no matter how low the probability, the most neglected, or that with the lowest cost? In an epistemic environment where so little is known, prioritising research direction in the way seemingly demanded by a 'well-ordered' science is, at the very least, challenging. Further, insofar as we are interested in extreme technological risk — extreme both in terms of possible impacts and (at least sometimes) in terms of low probabilities — systematic prioritisation becomes potentially unworkable. Either the extremity of the possible impact will flood the prioritisation with extreme risks or the minute probabilities will wash extreme risk out. We don't think all or most research done by science overall should be geared towards extreme risks, but neither should none of it.

A middle-ground approach is unlikely to be useful either, *in the context of this field of study*. An additional possible outcome to those noted above is to balance low likelihoods with extreme potential outcomes. However, this will probably yield a large collection of 'middling' topics to research. It is important to identify these, and many useful lines of enquiry can be pursued here, but they are not in the realm of existential risk which is, after all, by definition a study of extremes.

Moreover, deliberating over whether a particular line of enquiry is likely to be efficient/effective is equally difficult to parse, and again this is due to the epistemic nature of the field as we've described it. The only real means by which we can judge whether the field has 'achieved success' is by making note of whether societal collapse or some other irreversible outcome has occurred or been averted. Of course, this can only happen once, and not in a particularly useful or reproducible way as a data point. None of this is to say that considering questions of pursuit — what to research — for extreme risk is not critical to tackle, but instead to say that significant challenges face the field.

Given all the challenges that prioritisation in existential risk brings, we might be better served by aiming to understand whether the science is *well-adapted* (Currie, 2019a). This asks, given the investigations we want, whether the science is organised towards success; that is, are the incentives and resources governing scientific practice appropriate given the kind of knowledge we are after? There is no one rule for generating knowledge successfully, as we desire different kinds of knowledge and scientists face different epistemic situations in generating it. The epistemic situation of a science depends largely on the nature of the systems we wish to study and on the societal, technological, and ethical terrain the study is embedded within (Leonelli, 2016, Chapter 7; Currie, 2021). And for the reasons we have outlined above, existential risk studies face a peculiar epistemic situation: it considers unique events where there is little to no useful information available, it involves highly noisy complex systems with a great deal of interference, it must take into account second-order uncertainties, and so must encourage creativity (Currie, 2019a).

When we ask whether a scientific study of extreme technological risk is well-adapted, our focus shifts as we can have a clearer idea of the 'kinds' of lines of enquiry it would be fruitful for existential risk scholars to pursue: ones that are likely to lead to better futures for humanity. This is subtly but importantly different from the well-ordered framing above as it affords a more expansive view of what is worthwhile to study and, as we have seen above, and as Gero's work reminds us, this is an exploration of what is possible rather than uncovering truths, explicit probabilities, or a journey towards certainty. So, what this second framing asks is whether the social organisation that surrounds and supports the science is doing what it should to promote that exploration.

A common way of thinking about the social organisation of science examines the incentive structures scientists face (Currie, 2019b; Muldoon, 2013; Stanford, 2019). Scientists must secure positions and funding as well as publish and disseminate their findings, and doing this means overcoming various obstacles. Publishing and funding often require clear, testable hypotheses, passing peer review, having statistically significant results, and so on. We can understand these as various *standards* which shape the direction of scientific research. Such standards play multiple purposes simultaneously. On the one hand, they provide a common basis for scientific research to proceed: published results have been ascertained through community-sanctioned methods and meet similarly sanctioned criteria of significance and relevance. This facilitates communication and allows research to build upon itself. On the other hand, decisions scientists (and funders) make towards research direction are constrained by those standards. To an extent, highly constraining standards will lead to highly constrained science. A common complaint, for instance, is the insistence on publishing positive results, that is, results that are statistically significant, that reject the null, say. This means that the wide array of negative results do not make it onto the scientific record. And knowing when something doesn't work, or when effects are minimal, is surely knowledge as well.

So, one way of approaching questions of a science being well-adapted is to ask whether the standards are geared towards generating the kind of knowledge we should want or should expect from that science. We've argued that a science of extreme technological risk will be a science of possibility, of ambiguity, rather than a science of certainty. As such, many common features of scientific standards don't appear to lead to being well-adapted, as they are geared towards certainty. What, for instance, would count as a statistically significant result when we are dealing with a unique, or near-unique, event, which would have multiple wide-ranging effects across an extended temporal scale? For statistics to gain a grip, we need sufficient data and sufficient understanding of background conditions. For extreme technological risk, it seems, we plausibly have neither.

The challenge, then, is to develop a set of standards which are well-adapted for a science of ambiguity, and particularly for understanding such risk. Incentives should be geared towards understanding, and systematising, spaces of possibility: for understanding what is open, what is possible, what we can perhaps expect, given certain background conditions; for understanding what background conditions — particularly those we can control — plausibly lead us from catastrophic outcomes and towards safer outcomes, or even towards flourishing futures.

A well-adapted science of extreme technological risk, then, ought to develop standards which encourage this kind of open-ended exploration while nonetheless providing common grounds and baselines for research to accumulate.

2.4 Policy

Our previous challenge concerned, if you want, the internal organisation of science: what set of incentive structures and standards will facilitate generating the kind of knowledge we want? Our second concerns the influence of said science: how can a science of extreme technological risk not only articulate the factors influencing that risk but also aid in putting in place policies that lead us towards better futures? Again, the ambiguity of the science makes this a challenge.

There are a number of models that theorise the relationship between science and policy, and we'll approach the question of how a science of extreme technological risk should relate to policy through four of them.

Perhaps the most widespread — and perhaps the most naïve — model of the science–policy relationship claims that science tells us the probability of events: what the likely results are of adopting a particular policy (see Douglas, 2009, for objections). Policymakers take this knowledge into account, balance it against political and economic expediency, and decide what to do. The oft-repeated slogan 'following the science' appears to assume such a model; there is a one-directional epistemic arrow from science to policy. Of course, policy problems might stimulate research, but knowledge *flow* is still assumed to be unidirectional.

On such a model, policymakers want certainty or clear probabilities so they can ascertain the likely risk and benefits of various policy choices. We think generally this model is unfortunate: active scientific research rarely leads to true consensus, meaning that what 'following the science' looks like depends upon, well, which scientists you are following. Assessments of policy interactions have also found a great deal of selection bias in the types of evidence that are taken into consideration when setting policy (Boswell and Smith, 2017). Even as early as the 1970s, this linear model was found to be sorely lacking when it came to describing the actual relationship between science and policy in the traditional sciences (Weiss, 1979), but the model looks particularly unsuitable when we consider a science of ambiguity. Attempts to refine the 'knowledge → policy' model to make it more realistic often focus on the challenges of communication and the separateness of the communities involved. Policymakers and researchers don't speak the same language and so, while that flow of knowledge is ultimately what should be desired, what is lacking is effectiveness in how it is transmitted. However, even improved models fail to capture what is needed in a science of extreme technological risk: they are almost entirely geared towards establishing scientific certainty which is then (more or less effectively) communicated in a way that is able to inform policy. A science of ambiguity doesn't aim to inform policymakers about the probable outcomes of policies; instead, we inform them of the possibilities that are open given the adoption of said policies. There is no simple arrow between scientific results and policy.

A second model of the science–policy relationship posits that the arrow instead goes from politics to science: that any engagement with science will necessarily be in the service of a political project, whether this is broadly construed (such as in how 'expertise' is constructed, what research is legitimised) or in the particular (such as in research commissioned by interested groups to achieve or support certain policy outcomes). This is both a descriptive model, in the sense that it characterises the situation of science within a political context, and occasionally a more prescriptive model, for example, de Leeuw *et al.* (2014) used this to contend in the area of health research that '[researchers] would benefit from considering public policy through the lens of political science rather than through the lens of intervention research', while Nutbeam (2001) enjoins one 'to recognise the obvious political nature of the policy-making process and to engage more fully in that process'.

Either way, this can (perhaps obviously) have the result of stymying the creative potential of a science, with the model suggesting that under it, 'research that challenges dominant ideas will be discounted' (Boswell and Smith, 2017). One of us (Currie, 2019a) has argued that for reasons pertaining to the specific epistemic situation faced by existential risk studies, this discipline must be one that prioritises creativity but, as we have seen, this cannot flourish in a relationship wherein science services political projects. A policy relationship that follows an arrow from politics to knowledge thus cannot be the best environment for a useful science of extreme technological risk either.

A third approach from systems theory sees the science–policy relationship as consisting of two distinctly operating sociological spheres each of which selectively communicates to, and accepts communication from, the other. As Boswell and Smith (2017) describe it:

> On this account, science and politics are separate function-systems. Science (including social science) operates according to a binary code of true/false. In other words, it defines relevant communication based on whether it is concerned with establishing truth claims. The system of politics, meanwhile, selects relevant communication on the basis of the binary code of government/opposition. The political system selects and gives meaning to communication based on its relevance to the pursuit of political power and the capacity to adopt collectively binding decisions.

As a matter of descriptive accuracy, Boswell and Smith's encapsulation of this approach has much merit to it. However, as we have explored in this chapter, a science that deals almost entirely with extreme uncertainty is not one that is best served by 'establishing truth claims' even if this were possible. Nor does this model serve the governance side of the equation, given that it is only in the rarest of cases that politics deals with the timeframes that existential risk concerns itself with; there is rarely much benefit to be had in thinking in this way and therefore in governing by selecting and utilising this kind of science. Under this model, the relationship relies on each sphere 'picking up signals' (Boswell and Smith, 2017) from the other, often instrumentally, often selectively. But for the reasons just outlined, if the science and the politics truly do operate as separate (though communicating) spheres, existential risk studies will go nowhere in terms of pragmatic influence.

Whatever the directionality of the arrows, all of the models described thus far assume that science is in the business of abolishing uncertainty and making predictions; that is science's MO. If what we have said earlier holds water, this assumption won't do. A science of existential risk is one of ambiguity not certainty; such models then provide only limited insight for understanding just how the results and practices of such a science should inform policy or be communicated to policymakers.

A final, and we think potentially useful, model of the science–policy relationship recognises science's function in having 'conceptual influence' and that this influence can be incremental and slow. This conceptualisation of the use of science in policy is (oddly) also called 'enlightenment' (Weiss, 1977), whereby ideas drawn from scientific research percolate into policy environments without necessarily having a direct influence on any one policy at any one time. Rather, the policy environment is changed over time, gradually shifting as public and political views are shaped. This appears to lend itself rather well to existential risk: what needs to percolate is, essentially, the idea that there are many ways in which the future might unfold, and what needs instilling is a curiosity about (and wariness concerning) those futures. When framed as a piece of traditional policy advice, 'take the long term into consideration when making policy' seems rather toothless, vague, and unsatisfactory, but when seen as a constant drip-feed from multiple disciplines across months and years that gradually 'enlightens' those involved in policymaking, it could be quite powerful and more long-lasting in its impact. It also might help address some of the other policy challenges posed by the study of extreme technological risk, such as its complexity and its reliance on collaborations and input from multiple disciplines.

But this 'drip-feeding' model isn't really unidirectional, and to get a better idea of how, we turn to Science and Technology Studies, and a final model of science–policy interaction. Jasanoff has described the relationship between science and policy as one of co-production. According to her, '[s]cience in the co-productionist framework is understood as neither a simple reflection of the truth about nature nor an epiphenomenon of social and political interests' (Jasanoff, 2004). Instead, we have a complex bidirectional relationship: governance and policy are informed and shaped by science, but political forces shape research, too. This builds on the 'conceptual influence' idea described above but makes it explicitly two-way.

Importantly, under this framework, science and its processes are not seen as simply tools with which to solve problems. It is recognised — as in the study of extreme technological risk — that science and technology can be a potential source of problems for governance and policy to grapple with. Moreover, this organising theory of co-production has at its core four areas where co-production manifests, that we suggest in this chapter will be useful for existential risk studies to attend to: the emergence of new phenomena (for us, of ideas), disagreement surrounding those ideas, standardisation in how they are researched and promulgated, and enculturation in how research is performed.

Thus, a science of ambiguity that aims to influence policy — as we've argued the study of extreme technological risk should be — challenges simple models of the interaction between science and policy, leading us towards more complex, bi-directional approaches. How it is that such research should interact with policymakers and other stakeholders given these models remains an open challenge.

2.5 Conclusion

We've argued that the scientific study of extreme technological risk requires new ways of thinking about what we should expect from a science, how it should be organised, and how its relationships beyond science are conceived. It should be a science of ambiguity not certainty; possibility not actuality. As such, we shouldn't expect predictions, precise hypotheses, or probabilities. To be well-adapted, it should be incentivised towards understanding spaces of possibility and how these might be navigated. To be influential, such a science must interact with policymakers in new, perhaps more subtle, ways. Tackling extreme technological risk from a scientific perspective, then, is not simply a problem of how to generate and test hypotheses, and so on, but one that cuts deeply to the core of what we consider good science to look like, and what we think the relationship between science and society should be.

References

Avin, S., Wintle, B. C., Weitzdörfer, J., Ó hÉigeartaigh, S. S., Sutherland, W. J., and Rees, M. J. (2018). Classifying global catastrophic risks. *Futures*, 102, 20–26.

Beard, S. J., Rowe, T., and Fox, J. (2020). An analysis and evaluation of methods currently used to quantify the likelihood of existential hazards. *Futures*, 115, 102469.

Beard, S. J. and Torres, P. (2020). Ripples on the Great Sea of Life: A brief history of existential risk studies. Available at SSRN: https://papers.ssrn.com/sol3/papers.cfm?abstract_id=3730000.

Bostrom, N. (2013). Existential risk prevention as global priority. *Glob. Policy*, 4(1), 15–31.

Boswell, C. and Smith, K. (2017). Rethinking policy 'impact': Four models of research-policy relations. *Palgrave Commun.*, 4, 20.

Buskell, A. and Currie, A. (2021). Uniqueness in the life sciences: How did the elephant get its trunk? *Biol. Phils.*, 36(4), 1–24.

Cartwright, N. (2006). Well-ordered science: Evidence for use. *Philos. Sci.*, 73(5), 981–990.

Collingridge, D. (1980). *The Social Control of Technology*. UK: Frances Pinter (Publishers) Limited.

Currie, A. (2019a). Creativity, conservativeness & the social epistemology of science. *Stud. Hist. Philos. Sci.*, Part A, 76, 1–4.

Currie, A. (2019b). Existential risk, creativity & well-adapted science. *Stud. Hist. Philos. Sci., Part A*, 76, 39–48.

Currie, A. (2023). Science & speculation. *Erkenntnis*, 88, 597–619.

de Leeuw, E., Clavier, C., and Breton, E. (2014). Health policy — Why research it and how: Health political science. *Health Res. Policy Syst.*, 12(55).

Douglas, H. (2009). *Science, Policy, and the Value-free Ideal*. USA: University of Pittsburgh Press.

Genus, A. and Stirling, A. (2018). Collingridge and the dilemma of control: Towards responsible and accountable innovation. *Res. Policy*, 47(1), 61–69.

Gero, J. M. (2007). Honoring ambiguity/problematizing certitude. *J. Archaeol. Method Theory*, 14(3), 311–327.

Jasanoff, S. (2004). The idiom of co-production, Chapter 1. In S. Jasanoff (ed.), *States of Knowledge: The Co-production of Science and the Social Order* (1st edn.) (pp. 1–12). London: Routledge.

Kareiva, P. and Carranza, V. (2018). Existential risk due to ecosystem collapse: Nature strikes back. *Futures*, 102, 39–50.

Kitcher, P. (2003). *Science, Truth, and Democracy*. UK: Oxford University Press.

Leonelli, S. (2016). *Data-Centric Biology: A Philosophical Study*. Chicago, IL, USA: University of Chicago Press.

Liu, H. Y., Lauta, K. C., and Maas, M. M. (2018). Governing boring apocalypses: A new typology of existential vulnerabilities and exposures for existential risk research. *Futures*, 102, 6–19.

Manheim, D. (2020). The fragile world hypothesis: Complexity, fragility, and systemic existential risk. *Futures*, 122, 102570.

Millett, P. and Snyder-Beattie, A. (2017). Existential risk and cost-effective biosecurity. *Health Secur.*, 15(4), 373–383.

Muldoon, R. (2013). Diversity and the division of cognitive labor. *Philos. Compass*, 8(2), 117–125.

Nutbeam, D. (2001). Evidence-based public policy for health: Matching research to policy need. *UHPE Promot Education*, 2(Suppl), 5–27.

Ó hÉigeartaigh, S. (2017). The state of research in existential risk. In B. J. Garrick (ed.), *Catastrophic and Existential Risk: Proceedings of the First International Colloquium*, sponsored by The B. John Garrick Institute for the Risk Sciences, Luskin Convention Center, UCLA.

Pegram, T. and Kreienkamp, J. (2019). *Governing Complexity: Design Principles for Improving the Governance of Global Catastrophic Risks*. Global Governance Institute Policy Brief Series. London: University College London.

Sears, N. A. (2020). Existential security: Towards a security framework for the survival of humanity. *Glob. Policy*, 11(2), 255–266.

Stanford, P. K. (2019). Unconceived alternatives and conservatism in science: The impact of professionalization, peer-review, and Big Science. *Synthese*, 196(10), 3915–3932.

Weiss, C. (1977). Research for policy's sake: The enlightenment function of social research. *Policy Anal.*, 3(4), 531–545.

Weiss, C. (1979). The many meanings of research utilization. *Public Admin Rev.*, 39(5), 426–431.

Chapter 3

Risk and Scientific Reputation: Lessons from Cold Fusion

Huw Price

3.1 Introduction

Many scientists have expressed concerns about potential catastrophic risks associated with new technologies. But expressing concern is one thing, identifying serious candidates another. Such risks are likely to be novel, rare and difficult to study; data will be scarce, making speculation necessary. Scientists who raise such concerns may face disapproval not only as doomsayers but also for their unconventional views. Yet, the costs of false negatives in these cases — of wrongly dismissing warnings about catastrophic risks — are by definition very high. For these reasons, aspects of the methodology and culture of science, such as its attitude to epistemic risk and to unconventional views, are relevant to the challenges of managing extreme technological risk. In this piece, I discuss these issues with reference to a real-world example that shares many of the same features, that of so-called 'cold fusion'.

3.2 CSER and Maverick Science

Plans for the Cambridge Centre for the Study of Existential Risk (CSER) first emerged in conversations between Martin Rees, Jaan Tallinn, and me, with others, in Spring 2012. In one of those discussions, I remarked to Rees that some of the issues we wanted CSER to study had a poor reputation. They were regarded as 'a bit flaky', as I put it. Rees agreed, but said that this was why the project was important. Serious risks might not be getting the attention they deserved because of these reputational factors.

From that point, we were clear that a useful role for CSER might be to act as a reputational counterweight. In other words, CSER could use the reputation that we ourselves had at hand — that of Cambridge, and of our distinguished supporters and collaborators — as an opposing force to nudge these neglected issues away from the fringes towards respectability. In this way, we could help determine which neglected risks really needed attention and which could safely be left on the sidelines.

Three years later, when CSER won funding for the project *Managing Extreme Technological Risk* from the Templeton World Charity Foundation (TWCF), the role of reputation in science was an explicit focus of one of five subprojects. This is how we presented this work in our application to TWCF:

> *Extreme risk and the culture of science.* Prediction and mitigation of ETRs [extreme technological risks] is likely to depend on long-range evaluation of possibilities that seem far-fetched, in some cases. Many of these possibilities may turn out to be of negligible concern, but the net needs to be cast widely in the first instance, to maximise our chances of catching the fish that matter, as early as possible. Given the nature of the risks involved, there is a high cost to 'false negatives'.

> Unfortunately, science is not good at casting its net widely. As Kuhn (1962) observed, science is conservative, and there is strong cultural pressure on scientists to work within the current paradigm. Advances — Kuhn's 'scientific revolutions' — often depend on far-sighted individuals who resist these pressures, to work outside the mainstream. The history of science offers many examples of such figures, whose work is often shunned for long periods, before eventual vindication. Of course, history offers far more examples of fringe proposals that were not vindicated by later developments. In general, we rely on the normal process of science to sort out the gems from the dross — it may take a long time in some cases, but we get there in the end. In the special case of ETR, however, such a delay might be extremely costly. This subproject investigates this danger, and ways to reduce it.

Two years later again, in 2017, CSER organised a workshop on 'Risk and the Culture of Science'. It was held at Trinity College, Cambridge, in association with the When Experts Disagree (WEXD) Project based at University College, Dublin. We described the theme like this:

> Many scientists have expressed concern about potential catastrophic risks associated with powerful new technologies. But expressing concern is one thing, identifying serious candidates another. By definition, such risks will be novel, rare and difficult to study; data will be scarce, speculation necessary. This pushes us to the fringes of science, the realm of 'mavericks' and the unconventional — often a hostile and uncomfortable place. Scientists value consensus, at least about the big issues.
>
> Catastrophic risk is both a big issue and a highly charged one: so fringe-dwellers may be doubly unwelcome. Do we need to make special efforts to protect our mavericks, if catastrophic risk is to get the attention it deserves? If so, how can we do it? Can we use the values of science to protect useful fringe-dwellers from science's own immune system? Can we engineer a Maverick Room? (CSER, 2017)

The workshop involved a number of leading philosophers of science, including Professor Heather Douglas (Michigan State), whose work on the intersection of epistemic risk and value in science had been one of the inspirations for the project (Douglas, 2000, 2007, 2009). It also included some speakers we called our mavericks — researchers who felt that they had encountered these reputational issues in their own work. They spoke about their own experience in controversial fields such as nanotechnology risk, AI risk, geoengineering, and so-called 'cold fusion'.

The last example was particularly interesting, from my personal point of view. At around the same time as plans for CSER were first emerging, I had happened to become interested in claims then being made about cold fusion, or LENR ('low-energy nuclear reactions'), as it was also termed. By the time of the workshop in 2017, I had been following the field for several years. I regarded it as a fascinating real-world example of maverick science, in the sense we had in mind.

Like many millions of others, I had been aware of the claims of cold fusion after its public and controversial debut in 1989. I had kept an eye on it for some time afterwards, on the online forums provided in those days by internet newsgroups. But it had dropped off my radar for many years. Then, late in 2011, a new-generation online forum — Facebook — happened to make it visible to me again. As I'm going to explain, I have followed it ever since, writing several public pieces about it, and meeting some of the leading figures in the field, both inside and outside academia.

Nothing in those ten years has shaken my conviction that cold fusion is a fascinating real-world example of maverick science, in the sense relevant to the study of extreme technological risk. Indeed, I have come to see my own experience in thinking and writing about the field — in particular, some of the reactions I have encountered from others — as an interesting illustration of some of the general characteristics of maverick science. The present piece is a kind of ten-year progress report, from my personal and professional point of view — that of a philosopher of science, with an interest in the science of extreme risk. I'll tell the story of my own engagement with the field, and describe the lessons I think that we should take from it.

I'll tell this story, in part, by reproducing three of my public pieces, from 2015, 2017, and 2019 — they appear below as the starred Sections 3.3*, 3.5*, and 3.7*, respectively; readers familiar with these pieces may skip over them, of course. I'll fill out the narrative provided by these pieces with some additional detail about my engagement with the field at various points, and about developments in the field itself, especially in the years since 2019. I'll close with an assessment of where I think the field stands, and what lessons I think the case carries for risk and the culture of science.

My first public piece appeared in the online magazine *Aeon* in December 2015 (Price, 2015). The text below is from a lightly edited version I prepared for a WEXD workshop in Dublin in July 2017. Apart from three new introductory paragraphs — the first of them sadly apt in light of recent events — the main difference is that the *Aeon* version left out some of my original references and the Dublin version put these back in. I have tweaked the tenses in a few places, and added citations formerly provided by hyperlinks. And I have restored the title that I gave the piece originally in 2015 — *Aeon* preferred something a little less obscure! The Dublin version also included a postscript from 2017, with some updates about the field. It appears separately below (Section 3.4). Apart from that, I have resisted the temptation to update the piece with the benefit of hindsight.

3.3* My Dinner with Andrea (December 2015)

Science can be unkind to its mavericks — often they are shunned, even ridiculed. Sometimes, like the Australians Barry Marshall and Robin Warren (who eventually won a Nobel Prize for their discovery that stomach ulcers are caused by a bacterium), the mavericks get the last laugh. But should they have had to endure all those first laughs? We would have had better ulcer treatments sooner if Marshall and Warren had been listened to earlier, presumably. It is easy to imagine how the costs of delay might be much more severe. Suppose a maverick had identified some previously unnoticed pathway to a new pandemic — millions of lives might hang on the question of who was laughing at whom.

I'm a philosopher of science, and also co-founder of Cambridge's Centre for the Study of Existential Risk, so pessimistic speculations like these are part of my job. But in recent years I've been watching a real-life case. I've been following the fortunes of one of the most widely ridiculed groups in science (or pseudoscience, as their critics would say), the advocates of so-called 'cold fusion'. This community lives in what I now call a reputation trap, tainted with a reputation so dismissive, and so contagious, that the entire area is effectively off-limits to mainstream investigators. Anyone who is seen to approach the area with an open mind risks being pushed into the trap themselves — I speak from experience!

I'm increasingly convinced that the culture of reputation traps is unhealthy for science, and potentially dangerous for all of us, at least in cases where the costs of 'wrongful dismissal' are especially high. In particular, I'm worried that science itself is caught in a kind of meta trap. It is unable to consider the possibility that it is insufficiently open-minded, because the reputational fences themselves prevent it from asking whether the fences are too high or erected in the wrong places.

My own journey towards the fringes began in 2011, when a physicist friend made a joke on Facebook about the number of laws of physics being broken in Italy. He had two pieces of news in mind. One was a claim by the Gran Sasso-based OPERA team of having discovered superluminal neutrinos (Reich, 2011). The other concerned an engineer from Bologna called Andrea Rossi, who claimed to have a cold fusion reactor producing commercially useful amounts of heat (Hambling, 2011).

Why were these claims so improbable? The neutrinos challenged a fundamental principle of Einstein's theory of special relativity that nothing can travel faster than light, while cold fusion, or LENR, is the controversial idea popularised by Martin Fleischmann and Stanley Pons (1989) that nuclear reactions similar to those in the sun could also occur at or close to room temperature, under certain conditions. Fleischmann and Pons claimed to have found evidence that such reactions could occur in palladium loaded with deuterium (an isotope of hydrogen). A few other physicists, including Sergio Focardi at Bologna (Wikipedia contributors, 2021a), claimed similar effects with nickel and ordinary hydrogen. But most were highly sceptical, and the field 'subsequently gained a reputation as pathological science', as Wikipedia puts it (Wikipedia contributors, 2021b). Even the believers had not claimed commercially useful quantities of excess heat, as Rossi now reported from his 'E-Cat' reactor.

It turned out that my physicist friend and I disagreed about which of these unlikely claims was the less improbable — he thought the neutrinos, on the grounds that the work had been done by respectable scientists, rather than by a lone engineer with a somewhat chequered past, while I thought Rossi, on grounds of the physics. Superluminal neutrinos would overturn a fundamental tenet of relativity, but all Rossi needed was a previously unnoticed channel to a reservoir of energy whose existence is not in doubt. We know that huge amounts of energy are locked up in metastable nuclear configurations, trapped like water behind a dam. There's no known way to get useful access to that energy at low temperatures. But — so far as I knew — there was no 'watertight' argument that no such method exists.

My friend agreed with me about the physics (so has every other physicist I've asked about it since). But he still put more weight on the sociological factors — reputation, as it were. So, we agreed to bet a dinner on the issue. My friend would pay if Rossi turned out to have something genuine, and I would pay if the neutrinos came up trumps. We'd split the bill if, as then seemed highly likely, both claims turned out to be false.

It soon became clear that I wasn't going to lose. The neutrinos were scratched from the race, when it turned out that someone on OPERA's team of respectable scientists had failed to tighten an optical lead correctly (Reich, 2012). Rossi, however, seemed to be going from strength to strength.

While it is fair to say that the jury is still out (literally, as we'll see), there has been a lot of good news — for the field and my hopes of a free dinner — in the past few years. There have been two reports (Levi *et al.*, 2013, 2014) of tests of Rossi's device by teams of Swedish and Italian physicists whose scientific credentials are not in doubt, and who had access to one of his devices for extended periods (a month, for the second test). Both reports claimed levels of excess heat far beyond anything explicable in chemical terms, in the testers' view. (The second report also claimed isotopic shifts in the composition of the fuel.) Since then, there have been several reports of duplications by experimenters in Russia (Lewan, 2015a) and China (Wang, 2015), guided by details in the 2014 report.

More recently (by late 2015), Rossi was granted a US patent for one of his devices, previously refused on the grounds that insufficient evidence had been provided that the technique worked as claimed (Lewan, 2015b). There were credible reports that a 1-MW version of his device, producing many times the energy that it consumes, had been on trial in an industrial plant in Florida for months, with good results so far. And Rossi's US backer and licensee, Tom Darden — a respectable North Carolina-based industrialist, with a long track record of investment in pollution-reducing industries — has been increasingly willing to speak out in support of the LENR technology field (Dumaine, 2015).

Finally, there was a paper by two senior Swedish physicists, Rickard Lundin and Hans Lidgren, proposing a mechanism for Rossi's results, inspired in part by the second of the two test reports mentioned above (Lundin and Lidgren, 2015). Lundin and Lidgren say that the 'experimental results by Rossi and co-workers and their E-Cat reactor provide the best experimental verification' of the process they propose.

As I say, I don't claim that this evidence is conclusive, even collectively. It is still conceivable that there is fraud involved, as many sceptics have claimed, or some large and persistent measurement error. Yet, as David Bailey and Jonathan Borwein pointed out (Bailey and Borwein, 2014, 2015), these alternatives seemed to be becoming increasingly unlikely — which seemed great news for my dinner prospects.

Moreover, Rossi was not the only person claiming commercially relevant results from LENR. Another prominent example is Robert Godes, of Brillouin Energy, profiled in a recent Norwegian newspaper piece (Bjørkeng, 2015). If you want to dismiss Rossi on the grounds that he's claiming something impossible, one of these explanations (i.e., fraud or large and persistent error) needs to work for Godes, too.

You can see why I've been salivating at the thought of My Dinner with Andrea, as I've been calling it (Malle, 1981), in honour of the man who would be the absent guest of honour, if my physicist friend is paying. And it is not only my stomach that has been becoming increasingly engaged with this fascinating story. I'm a philosopher of science, and my brain has been finding it engrossing, too.

What do I mean? Well, it hasn't escaped my attention that there's a lot more than a free dinner at stake. Imagine that someone had a working hot fusion reactor in Florida — assembled, as Rossi's 1-MW device is reported to be, in a couple of shipping containers, and producing several hundred kilowatts of excess power, month after month, in apparent safety. That would be huge news, obviously. (As several people have noticed, a new clean source of energy would be really, really useful, right about now!)

But if the potential news is this big, why haven't most of you heard about Rossi, or Godes, or any of the other people who have been working in the area (for many years, in some cases)? This is where things get interesting, from a philosopher of science's point of view.

As a question about sociology, the answer is obvious. Cold fusion is dismissed as pseudoscience, the kind of thing that respectable scientists and science journalists simply don't talk about (unless to remind us of its disgrace). As a recent *Fortune* piece puts it, the Fleischmann and Pons 'experiment was eventually debunked and since then the term cold fusion has become almost synonymous with scientific chicanery' (Dumaine, 2015). In this case, the author of the article is blithely reproducing the orthodox view, even in the lead-in to his interview with Tom Darden — who tells him a completely different story (and has certainly put his money where his mouth is).

Ever since 1989, in fact, the whole subject has been largely off-limits, in mainstream scientific circles and the scientific media. Authors who do put their head above the parapet are ignored or rebuked. Most recently, Lundin and Lidgren report (Lewan, 2015c) that they submitted their paper to the journal *Plasma Physics and Controlled Fusion*, but that the editors declined to have it reviewed, and that even the non-reviewed pre-print archive, arxiv.org, refused to accept it.

So, as a matter of sociology, it is easy to see why Rossi gets little serious attention; why an interview with Tom Darden associates him with scientific chicanery; and why, I hope, some of you are having doubts about me for writing about the subject in a way that indicates that I am prepared to consider it seriously. (If so, hold that attitude; I want to explain why I take it to reflect a pathology in our present version of the scientific method. My task will be easier if you are still suffering from the symptoms.)

Sociology is one thing, but rational explanation another. It is very hard to extract from this history any satisfactory *justification* for ignoring recent work on LENR. After all, the standard line is that the rejection of cold fusion in 1989 turned on the failure to replicate the claims of Fleischmann and Pons. Yet, if that were the real reason, then the rejection would have to be provisional. Failure to replicate couldn't possibly be more than provisional — empirical science is a fallible business, as any good scientist would acknowledge. In that case, well-done results claiming to overturn the failure to replicate would certainly be of great interest.

Perhaps the failure to replicate wasn't crucial after all? Perhaps we knew on theoretical grounds alone that cold fusion was impossible? But this would make nonsense of the fuss made at the time and since, about the failure to reproduce the Fleischmann and Pons results. And in any case, it is simply not true. As I said at the beginning, what physicists actually say (in my experience) is that although LENR is highly unlikely, we cannot say that it is impossible. We know that the energy is in there, after all.

No doubt, one could find some physicists who would claim it was impossible. But they might like to recall the case of Lord Rutherford, the greatest nuclear physicist of his day, who famously claimed that 'anyone who expects a source of power from transformation of … atoms is talking moonshine' — the very day before Leo Szilard, prompted by newspaper reports of Rutherford's remarks, figured out the principles of the chain reaction that makes nuclear fission useable as an energy source, peaceful or otherwise (The story is told in Rhodes, 1986, Chapter 1).

This is not to deny that there is truth in the principle popularised by Carl Sagan that extraordinary claims require extraordinary evidence. We should certainly be very cautious about such surprising claims, unless and until we amass a great deal of evidence. But this is not a good reason for ignoring such evidence in the first place, or refusing to contemplate the possibility that it might exist. As Robert Godes said recently, 'It is sad that such people say that science should be driven by data and results, but at the same time refuse to look at the actual results' (Bjørkeng, 2015).

Again, there's a sociological explanation for why few people are willing to look at the evidence. They put their reputations at risk by doing so. Cold fusion is tainted, and the taint is contagious; anyone seen to take it seriously risks contamination themselves. So the subject is stuck in a place that is largely inaccessible to reason — a reputation trap, we might call it. People outside the trap won't go near it, for fear of falling in. 'If there is something scientists fear it is to become like pariahs', as Rickard Lundin puts it (Lewan, 2015c). People inside the trap are already regarded as disreputable, an attitude that trumps any efforts they might make to argue their way out, by reason and evidence.

Outsiders might be surprised how well populated the trap actually is in the case of cold fusion and LENR. The field never entirely went away, nor vanished from the laboratories of respected institutions. Rossi's own background is not in these laboratories, but he acknowledges that his methods owe much to those who are, or were — especially to the late Sergio Focardi, one of the pioneers of the field (Wikipedia Contributors, 2021a). To anyone willing to listen, the community will say that they have amassed a great deal of evidence of excess heat, not explicable in chemical terms, and of various markers of nuclear processes. Some, including a team at one of Italy's leading research centres, say that they have many replications of the Fleischmann and Pons results (ENEA, 2013).[1]

---

[1] I was delighted to discover recently that some of the early Italian cold fusion research took place in the very same underground laboratory as the 2011 work that (briefly) reported superluminal neutrinos. The story is told in a fascinating 1994 essay by the Caltech physicist David Goodstein, who had professional colleagues in both camps, in the heated early controversy about cold fusion. Here, he describes some of the work of the ENEA team led by his friend Professor Francesco Scaramuzzi:

> Reacting to criticism of the primitive technique they had used to detect neutrons, they purchased the best neutron detection system in the world, essentially identical to the one used by Charlie Barnes at Caltech. Going one better, they installed it in physics laboratories that had been excavated under a mountain called the Gran Sasso, a two-hour drive from Rome. Anywhere on the surface of the Earth, there are always some neutrons buzzing around due to cosmic radiation from outer space. This so-called "background" has to be subtracted from the neutrons produced by any other phenomenon such as Cold Fusion. In the galleries under the Gran Sasso, the shielding effect of the mountain reduces the cosmic ray neutron background nearly to zero. That's why the laboratory was built there. An automated system was set up to monitor the neutron counter while running the temperature of a Scaramuzzi-type deuterium gas cell up and down. Every week or so, a member of the group would have to drive out to the Gran Sasso lab, check out the counters, replenish the supply of liquid nitrogen, and bring back the data. No one could accuse them any longer of being unsophisticated about neutron work. However, this experiment, like their own earlier work and many others blossoming around the world, produced positive results, but only sporadically. There was no dependable recipe for coaxing bursts of neutrons out of the Cold Fusion cell (Goodstein, 1994).

Again, the explanation for ignoring these claims cannot be that other attempts failed twenty-five years ago. That makes no sense at all. Rather, it is the reputation trap. The results are ignored because they concern cold fusion, which we 'know' to be pseudoscience — we know it because attempts to replicate these experiments failed, twenty-five years ago! The reasoning is still entirely circular, but the reputation trap gives its conclusion a convincing mask of respectability. That's how the trap works.

Fifty years ago, Thomas Kuhn (1962) taught us that this is the usual way for science to deal with paradigm-threatening anomalies. The borders of dominant paradigms are often protected by reputation traps, which deter all but the most reckless or brilliant critics. If LENR were an ordinary piece of science (or proposed science), the challenge by Rossi and others would provide some fascinating spectator sport for philosophers and historians of science — a Kuhnian revolution waiting to happen, perhaps, with threats to the stability of the reputation trap now clearly in view. We could take our seats on the sidelines, and wait to see whether walls fall — whether distinguished sceptics end up with egg on their faces. 'Pleasant is it to behold great encounters of warfare arrayed over the plains, with no part of yours in the peril', as Lucretius (1916) put it.

This would be plenty to explain why I've been finding Rossi's apparent progress so engaging. But there's more; much more. None of us, even philosophers, are mere spectators in this case. We all have skin in the game, and parts, indeed a planet, quite seriously in the peril. We are like a thirsty town, desperate for a new water supply. What we drink now is slowly killing us. We know that there's an abundant supply of clean, cheap water, trapped behind the dam. The problem is to find a way to tap it. A couple of engineers thought that they had found a way, twenty-five years ago, but they couldn't make it work reliably, and the profession turned against them. Since then, there's been a big reputation cost to any engineer who takes up the issue.

Put this way, it is easy to see an argument that we've been shooting ourselves in the foot. In a case like this, there is very little cost to a false positive — to investing some time and money in an avenue that turns out eventually to go nowhere. But there may be a huge cost to a false negative. If Rossi, Godes, Lundin, Lidgren, and others do turn out to have something useful — something that can make some useful contribution to meeting our desperate need for clean, cheap energy — we will have wasted a generation of progress, by dumping cold fusion in the reputation trap. What we should have done instead is to have engineered the exact opposite of a reputation trap — an X Prize-like reward for the first reliable replication of the Fleischmann and Pons results, above some commercial bar, perhaps.

Now, I can explain what I meant earlier, when I asked you to hold on to the thought that I must be a bit flaky myself, if that's your reaction to my willingness to take cold fusion seriously. If you do think that — at least if you think it without having studied the evidence for yourself — then your reaction is a symptom of the reputation trap. But I've now suggested that the trap itself may be an irrational pathology, in a special case like this, where the cost of a false negative is very high. If I'm right, then in a more rational world, we would fix our scientific norms to escape it. In a more rational world, you wouldn't think I'm flaky.

I don't need to deny that your reaction is an appropriate one, by the standards of science as we currently practice it, or that those standards work pretty well, in general. Reputation traps have a useful purpose, in the Kuhnian picture. They help to maintain the stability important to what Kuhn called normal science — the ordinary, useful kind of science when paradigms are not under threat. But this is compatible with the claim that they can be harmful in special cases (of which cold fusion may be one) — and that we could do better, if we were better at identifying those cases in advance.

I suspect it is too late to dismantle the trap for LENR — the horse is already in the process of bolting, I think. If so, then the field is going to be mainstream soon, in any case. But we could try to learn from our mistakes. There may be other potential cases with a similar payoff structure (a high cost for false negatives, with a low cost for false positives). I suspect there are some in the area of emerging extreme risks, another field in which I have some interest. I've met scientists employed as technology forecasters for a very large national enterprise who were frustrated at their inability to persuade their organisation to list LENR as a technology of possible strategic interest, even with low probability. Once again, it was the reputation trap at work. This attitude could be disastrous in other cases, if the ideas stuck in the trap are the key to early detection of some potentially catastrophic risks. (Indeed, as my physicist friend rightly points out, it might be disastrous in the case of LENR itself, if it turns out to generate its own extreme risks.)

There are big issues here. Would it be possible to avoid these counter-productive cases of the reputation trap, without erring too far in the other direction, without opening the floodgates, so to speak? I'm not sure, but I think it is important to put the question on the table. If LENR does develop in the direction I now think likely, we might be able to salvage something useful from mistakes in this case.

I'll close with some words from Tom Darden, from a speech at ICCF-19, the international meeting of the LENR community held in Padua, Italy, in April 2015. (A full transcript of the speech is accessible at McDonough, 2015.) If the field is indeed in the process of digging itself out of the repu-tation trap, then Darden deserves much of the credit. Either way, his atti-tude displays the kind of cautious open-mindedness that has been so lacking in reactions to the field for most of its history.

Darden describes how he came to invest in LENR. Until quite recently, he says, he shared the conventional view that 'the subject was dead'. But several independent enquiries about LENR within a matter of weeks convinced him that there was something worth investigating:

> We believed LENR technology was worth pursuing, even if we turn out to be unsuccessful ultimately. We were willing to invest time and resources to see if this might be an area of useful research in our quest to eliminate pollution. At the time, we were not especially optimistic, but the global benefits were compelling.

He reports that things have been going very well. 'We've had some success, and we're expanding our work ... and believe that we may be, at last, on the verge of a new paradigm shift'. Finally, he expresses his appre-ciation to his audience, the LENR community themselves, and recalls another controversial piece of Italian science:

> I would like to say how truly sorry I am that society has attacked you for the last three decades. The treatment of Fleischmann and Pons, and the treatment of many of you, by mainstream institutions and the media will go down in history as one more example of scientific infanticide, where entrenched interests kill off their divergent progeny. This seems to be a dark component of human nature, and I note the irony of it — we are in Padova, Galileo's home.

It would be easy to overstate the analogy between mainstream institutions and the Inquisition, but it isn't entirely empty. If we refuse to acknowledge the possibility that existing scientific institutions are not working as well as they might, we do something to reinforce it. If the reputation trap makes it impossible to question the role of the reputation trap, then the cardinals are winning.

3.4 The View from 2016

The above piece appeared in November 2015. I published a short update in *Aeon* in March 2016 (Price, 2016), when I was even more struck by the lack of attention that the Rossi saga was getting, as the end of the claimed one-year test of his reactor approached. In my view, it was a fascinating story whatever the outcome. Even if it was fraud, it was an extraordinarily elaborate and long-lasting fraud, in which the victims included some well-known individuals and financial institutions. Yet, there was virtually no coverage of the story, even from this angle. I took this to be a further manifestation of the reputation trap. One couldn't write about the story without putting on the table the view obviously held by Tom Darden and other financial backers that LENR might not be pseudoscience after all — and that view was taboo by the standards of the reputation trap.

More in a moment about what became of this part of the story. As I noted earlier, I updated my original *Aeon* article in May 2017. At that point, I had two new audiences in mind. The first, as I mentioned, was a workshop organised by the When Experts Disagree (WEXD) Project at University College, Dublin, in July 2017 (McKay, 2017); WEXD had been partners in our own 'mavericks' workshop at Trinity College, Cambridge, in April 2017. I'll come back to the second audience later. The new version included the following postscript, which begins with a brief update about Rossi's case.

3.5* Postscript (May 2017)

It has now been more than a year since I wrote about these ideas in *Aeon*. How has LENR been faring?

About Andrea Rossi's E-Cat, I said above that the jury was still out. That's now become literally true! The rumoured one-year test of Rossi's 1-MW reactor was completed in Spring 2016. As predicted, the verdict of the appointed 'expert responsible for validation' (ERV) was strongly positive. The ERV reported heat outputs many times the input energy, throughout the course of the test. However, Rossi's financial backers, Tom Darden's company Industrial Heat (IH), were not convinced. They declined to pay the $89m success fee to which Rossi claims they were contractually committed. Rossi took them to court, and that is where things stand.

The case comes before a jury in Florida in June 2017, and the pre-trial discovery process has already revealed a fascinating mass of information about Rossi's plant, about his history with IH, and about their interest in LENR in general. IH want to distance themselves from Rossi, but show no sign of regretting their commitment to the field as a whole. On the contrary, they appear to be seeking to accumulate as much as possible of the potential IP — as any rational investor would, presumably, if convinced that there was something there.

Elsewhere in the LENR field, there have been several interesting new announcements of positive results. One of the most striking is a report by the respected lab SRI International, based in Menlo Park, California, of reactors provided by Robert Godes's Brillouin Energy. Brillouin Energy was the second of the two claimed LENR developers I mentioned in my original article. The SRI tester, Dr Francis Tanzella, summarises his results as follows:

> We report here on the most recent nine months of extensive testing in Brillouin's two original [reactors] operated at its Berkeley laboratory, and in the past two months, with the second unit having been further situated at SRI. Brillouin has manufactured five identical metallic cores and has consecutively tested each one of them in its two [reactors], seemingly producing the same controlled heat outputs repeatedly.

> Since its reconstruction and calibration, I have been able to corroborate that the [reactor] system moved to SRI continues to produce similar LENR Reaction Heat that it produced up in its Berkeley laboratory at Brillouin. Together with my prior data review, it is now clear that these very similar results are independent of the system's location (Berkeley or Menlo Park) or operator (Brillouin's or SRI's personnel). This transportable and reproducible reactor system is extremely important and extremely rare. These two characteristics, coupled with the ability to start and stop the reaction at will are, to my knowledge, unique in the LENR field to date. (Tanzella, 2016, p. 2)

But this SRI report is not the only recent claim of reliable and reproducible results. Another came from the Condensed Matter Nuclear Science (CMNS) Department at Tohoku University, Japan, a program supported by the Japanese Clean Planet initiative. They, too, report 'highly reproducible' excess heat (Kaneko, 2016).

The Clean Planet initiative was established in 2012, after the Fukushima disaster the previous year, and is an enthusiastic backer of research on LENR. The reputation trap seems always to have been shallower in Japan, where there is a long history of LENR research by groups including Mitsubishi Heavy Industry Ltd., and what there was of it seems to have been swept away by the tragic events of 2011.[2]

In the West, it is still a different matter, but there are a few new cracks in the wall. In September 2016, *New Scientist* published an article entitled 'Cold fusion: Science's most controversial technology is back', reporting a surge of commercial interest in LENR:

> You won't hear the words "cold fusion", but substantial sums of money are quietly pouring into a field now known as low-energy nuclear reactions, or LENRs. Earlier this year, the US House of Representatives Committee on Armed Services declared it was "aware of recent positive developments" in developing LENRs and noted their potential to "produce ultra-clean, low-cost renewable energy" and their "strong national security implications". Highlighting too the interest of Russia, China, Israel and India, it suggested the US could not afford to be left behind, and requested that the Secretary of Defense provide a briefing. (Brooks, 2016)

As if to cover its reputational bases, *New Scientist* accompanied this piece with an unfriendly editorial, dismissing the suggestion that there is now a case for public funding of LENR research:

> Taxpayer money could provide credibility and ensure that results are properly reported, rather than just rumoured. And if there is anything to cold fusion, it would be in the public interest for it to be investigated properly. But that's an enormous if. There's still no compelling reason to

---

[2]Professor Jirohta Kasagi (Tohoku University) has recently informed me that in his view, the reputation trap has not been appreciably shallower in Japan. He surveys the early history of Japanese cold fusion research in Kasagi and Iwamura (2008).

think cold fusion will work. Let those with money to burn take the risk and, if proven right, the rewards for their chutzpah too. For the rest of us, cold fusion is better off left out in the cold. (Anonymous, 2016)

In my view, this editorial is a sad example of the irrational pathology of the reputation trap. Why? Well, imagine an analogy (Scott, 2015). You are marooned on Mars when the rest of the expedition leaves without you. Food is running short, but you could improve your chances by growing your own. Unfortunately, the only available source of fertiliser is the packaged waste from your former teammates — the crap they gave you before they left, as it were. There's 'no compelling reason' to think that this malodorous option will actually save you, but it is the best option in a tight corner. You hold your nose and get to work.

Let's agree with *New Scientist* that there's still no compelling reason to think cold fusion *will* solve our energy crisis. That's a very different matter from having a compelling reason to think that it *won't*. Our planet is in a very, very tight corner. How high does the probability need to be, before it makes sense to hold our reputational noses and join the search?

If there's one important difference between these two cases, it is that the stench of the reputation trap is an artefact of the culture of science itself, not an inevitable product of the biophysics of human byproducts. If we agreed to take LENR seriously, the odour would simply vanish. A better analogy would be with the castaway who starves to death on a fertile island because hunting and gathering are taboo for a man of his high caste. As in that case, the obstacle to doing the rational thing is one we have created for ourselves.[3]

When *New Scientist* says that 'not many respectable scientists would touch cold fusion', it is playing its own role in the cultural process that maintains the obstacle, adding a new small smear to the opprobrium with which cold fusion is coated. The resulting odour keeps us all in a trap, not only by preventing us from exploring a significant part of a

---

[3] In the light of my use of the Martian example, and my remark that the reputation trap seemed to be shallower in Japan, I was delighted to learn recently that 'in eighteenth century Japan, biosolids were an esteemed substance'. Due to its importance as a fertiliser, 'Japanese citizens did not view human waste as unwanted muck, but rather as something of value' (Zeldovich, 2019). So, there is less difference between the Martian case and the high caste case than I imagined.

desperately needed solution space but also — at what philosophers call the meta level — by discouraging us even from raising the possibility that such a prohibition might be deeply irrational.

3.6 Meeting Some of the Mavericks (2017–2018)

As I stated, I wrote the updated version of my original *Aeon* piece with two audiences in mind. One was the WEXD workshop at University College, Dublin, in July 2017. The second was proposed by the Australian science journalist, Mikey Slezak (a former student of mine at the University of Sydney). Slezak suggested that something based on my *Aeon* piece be included in the 2017 edition of an annual collection of Australian science writing. I gave him the updated version, with the postscript. In what I took to be yet another illustration of the reputation trap at work, it was then vetoed by a physicist on the committee, and never reached that audience.

A big benefit of the original *Aeon* piece, from my perspective, was that it brought me to the attention of many people within the LENR field. It also caught the eye of Clive Cookson, the *Financial Times* science editor who, as a young journalist, had written the very first newspaper story about the Fleischmann and Pons announcement in 1989. As Cookson (2012) recounts, he knew Martin Fleischmann at that point because his father had been a colleague of Fleischmann's in Chemistry at the University of Southampton. Fleischmann sought his advice, as a journalist and family friend, before the famous news conference — which Cookson advised him against.

Within the field, one of my initial contacts was with Alan Smith, a very active and well-connected UK-based researcher. Smith was the cold fusion maverick at our workshop in Cambridge in 2017. He introduced me to others in the field, including some of the leading Japanese researchers, whom I visited in Sendai in October 2017. They showed me some of their experiments, including cells that they said were reliably producing a few watts of excess heat for weeks at a time.

I also visited Stockholm, in November 2017, for what was billed as a demonstration of the latest version of Rossi's E-Cat. I met Andrea Rossi himself, and some of the Swedish and Italian scientists who had taken an interest in his work. Earlier in 2017, a few weeks after my postscript was written, Rossi had withdrawn his case against Industrial Heat, in an out-of-court settlement. Certainly, what had emerged in the legal process before that did not inspire great confidence that he had what he claimed, nor, in my view, did the Stockholm demonstration.

Let's now pick up the story with my second public piece (Price, 2019). This was written with an eye to the thirtieth anniversary of the infamous Fleischmann and Pons news conference, which fell in March 2019. By the time I wrote this piece, I had been introduced by email to the SRI researcher mentioned above, Francis Tanzella, whose private comments reinforced the impression I had about Brillouin's progress from public material. I had also been introduced to one of the major figures of the LENR world, the former SRI team leader, Dr Michael McKubre. Again, I have edited this piece as lightly as possible, mainly to add references.

3.7* Icebergs in the Room? Cold Fusion at 30 (March 2019)

From aviation to zookeeping, there's a simple rule for safety in potentially hazardous pursuits. Always keep an eye on the ways that things could go badly wrong, even if they seem unlikely. The more disastrous a potential failure, the more improbable it needs to be before we can safely ignore it. Think icebergs and frozen O-rings (Wikipedia Contributors, 2021c). History is full of examples of the costs of getting this wrong.

Sometimes, the disaster is missing something good, not meeting something bad. For hungry sailors, missing a passing island can be just as deadly as hitting an iceberg. So, the same principle of prudence applies. The more we need something, the more important it is to explore places we might find it, even if they seem improbable.

We desperately need some new alternatives to fossil fuels. To meet growing demands for energy, with some chance of avoiding catastrophic climate change, the world needs what Bill Gates called an energy miracle — a new carbon-free source of energy, from some unexpected direction. In this case, it is obvious what the principle of prudence tells us. We should keep a sharp eye out, even in unlikely corners.

Yet, there's one possibility that has been in plain sight for thirty years, but remains resolutely ignored by mainstream science. It is the so-called *cold fusion*, or *LENR* (for Low-Energy Nuclear Reactions). Cold fusion was made famous, or some would say infamous, by the work of Martin

Fleischmann and Stanley Pons in March 1989. Fleischmann and Pons claimed to detect excess heat at levels far above anything attributable to chemical processes in experiments involving the metal palladium, loaded with hydrogen. They concluded that it must be caused by a nuclear process — 'cold fusion', as they termed it.

Many laboratories failed to replicate Fleischmann and Pons's results, and the mainstream view since then has been that cold fusion was 'debunked'. It is often treated as a classic example of disreputable pseudoscience. But it never went away completely. It has always had defenders, including some scientists at very respectable laboratories. They acknowledged that replication and reproducibility were difficult in this field, but claimed that most attempts on which the initial dismissal had been based were simply too hasty.

Such work continues today, as cold fusion approaches its thirtieth birthday. A recent peer-reviewed Japanese paper lists seventeen scientific authors from several major universities and the research division of Nissan Motors (Kitamura *et al.*, 2018). These authors report 'excess heat energy', which 'is impossible to attribute … to any chemical reaction' (with good reproducibility between different laboratories). The field has also been attracting new investors recently (including, some claim, Bill Gates himself).

These 17 Japanese scientists might be mistaken, of course. Scientists — not to mention investors — often get things wrong. But their work is only the tip of a very substantial iceberg. If there was even a small probability that they and the rest of the iceberg were on to something, wouldn't the field deserve some serious attention, by the prudence principle with which we began?

When I wrote about these issues in *Aeon* 3 years ago, I argued that the problem is that cold fusion is stuck in a reputation trap. Its image is so bad that many scientists feel that they risk their own reputations if they are seen to be open-minded about it, let alone supporting it. That's why the work of those Japanese scientists and others like them is ignored by mainstream science — and why it doesn't get the attention that simple prudence recommends.

The reputation trap is nicely illustrated by the tone of a *New Scientist* editorial from 2016. It accompanied a fairly even-handed article (Brooks, 2016) describing recent increases in interest in LENR, from investors as well as some scientists. The editorial concludes:

There's still no compelling reason to think cold fusion will work. Let those with money to burn take the risk and, if proven right, the rewards for their chutzpah too. For the rest of us, cold fusion is better off left out in the cold. (Anonymous, 2016)

There's no mistaking the tone, but if we translate it to the safety case, the logic has a chilling familiarity: 'There's no compelling reason to think that there will be icebergs at this latitude. Let those with money to burn take the slower route to the south, and the rewards if they turn out to be right'.

The fallacy here is obvious. It puts the burden of proof on the wrong side. What matters is not whether there is a compelling reason to think that there *are* icebergs, but whether there is compelling reason to be confident that there *are not.* That's what's distinctive about these safety cases, and it stems from the high cost of getting things wrong — hitting the icebergs, or missing the islands.

In the safety case, we know what happens when reputation and similar cultural and psychological factors get in the way of prudence. Icebergs are unlikely, and our reputation is at stake, so full speed ahead! NASA fell for precisely this trap in the case of the *Challenger* disaster, ignoring warnings about the O-rings (Berkes, 2012). Something similar underlies the tone of the *New Scientist* editorial, in my view — a kind of misplaced rigidity, engendered in this case by the norms of scientific reputation.

Reputation plays an indispensable role in science, as an aid to quality control. But sometimes it gets things wrong. There are famous cases in the history of science in which new ideas were ignored or ridiculed, sometimes for decades, before going on to win Nobel Prizes. (Classic examples include the work of Barbara McClintock on mobile elements in genetic material and the discovery by Australian scientists Barry Marshall and Robin Warren that stomach ulcers are caused by a bacterium.)

Usually this doesn't matter very much — science got there in the end, in these famous cases. But it is easy to see how it might be a problem, where prudence requires that we take unlikely possibilities very seriously. If what's at stake is a serious risk, the normal rate of progress in science — one funeral at a time, as Max Planck put it, commenting on science's conservatism — might simply be too slow.

So, the normal sociology of scientific reputation may be pathological in special cases — cases in which the cost of wrongly dismissing a maverick idea is especially high. In my *Aeon* piece, I suggested that LENR is such a case. I proposed that to offset this pathology we need some carefully targeted incentives — an X Prize for new energy technologies, say. To mainstream scientists, this idea sounds absurd, even disreputable, at least in the case of cold fusion. But that's just the pathology talking, in my view — and the rational response to the pathology is to hack it and work around it, not to give way to it.

Not surprisingly, my article was controversial — some commentators wondered what it would do to my own reputation! Critics didn't disagree with the principle that we need to take low-probability risks (or potential missed opportunities) seriously when the cost of overlooking them would be high. But many denied that cold fusion falls into this category. They felt that it is so unlikely, so discreditable, that we can safely leave it in the reputation trap. Sometimes, this response came with considerable vehemence, even from friends.

How likely would cold fusion have to be, to be worth serious attention? This is debatable, but a generous 5% should be uncontroversial. (Who would argue that we should ignore a 1 in 20 chance of some interesting new physics, let alone carbon-free energy?) My critics thought that the probability that cold fusion is real is much lower than that.

I felt that many of these critics were simply not paying attention. If one took the trouble to look, there was a lot of serious work, including recent work, suggesting real physical anomalies. If we ask not whether this evidence is entirely compelling, but simply whether it lifts the field above a very low attention threshold (say 5%), the answer seemed to me to be obvious. We shouldn't be ignoring this work. Instead, we should be trying to hack the pathology that makes it so easy to dismiss it.

In addition to scientists at respectable institutions who work on LENR, there are also some inventors and entrepreneurs who claim to be developing practical LENR-based devices. I mentioned two in my 2015 article. One was a controversial Italian engineer, Andrea Rossi. His claims in 2011 had attracted me to the topic in the first place, and in 2015 he seemed to be doing well. The other was a less colourful inventor, Robert Godes, whose Berkeley-based company Brillouin Energy also claimed to be on a path to a commercial LENR reactor.

My critics were confident that both Mr. Rossi and Mr. Godes must be frauds, or else deeply confused. What other possibilities are there, after all, if — as my critics were convinced — there's no genuine LENR? I thought that this dismissal was far too hasty. I wasn't certain by any means that Rossi or Godes did have what they claimed, but I thought that the probability was well above a reasonable attention threshold (given what success might mean).

With several critics, these differences of opinion led to bets, at long odds. I would win the bets if, after three years, either Rossi or Godes had 'produced fairly convincing evidence (>50% credence) that their new technology generates substantial excess heat relative to electrical and chemical inputs'. If my opponents and I couldn't agree whether this is the case, the question would go to a panel of three judges for arbitration. Either way, the proceeds will support research on existential risk.

The three years is now up, so how am I faring? About Rossi, I am happy to concede that he hasn't made it to the finishing line, even at a modest 50% credence. I think there is still some reason to think that he may have *something*, based in part on claimed replications by far less colourful figures. But there is also evidence of dishonesty, especially in his dealings with his US backer, Industrial Heat.

Luckily for me, I backed the ants as well as the grasshopper. About Godes's Brillouin Energy (BEC), the news is much better. There are now three positive reports by an independent tester, Dr Francis Tanzella, at the Menlo Park lab of SRI International (Tanzella, 2016, 2018a, 2018b). The first report already confirmed low levels of excess heat, and important progress in reproducibility:

> This transportable and reproducible reactor system is extremely impor-
> tant and extremely rare. These two characteristics, coupled with the
> ability to start and stop the reaction at will are, to my knowledge, unique
> in the LENR field to date.

The more recent reports describe steady progress in two directions: first, a modest improvement in excess heat as measured by the so-called Coefficient of Performance (COP) — the ratio of output power to input power; second, a large increase in the *absolute* level of excess heat, from a few milliwatts in 2016 to several watts in early 2018.

The last of Dr Tanzella's three reports covers the period to July 2018. Since then, BEC themselves have claimed even better results — consistent output power around twice the level of input power, with excess heat of around 50 watts.

What are the options, if we are not to take these reports at face value? Essentially, one needs to dismiss as incompetent or fraudulent not only

Mr Godes and his BEC team, but also Dr Tanzella and his SRI colleagues. However, as the 2016 report notes, SRI 'brought over 75 person-years of calorimeter design, operation, and analysis experience to this process' (Tanzella, 2016, p. 1), much of it in the field of LENR. SRI and Dr Tanzella himself are among the most experienced experts in the field.

Accordingly, it seems to me greatly more likely than not that BEC do have what they claim — in the words of my bet, a device that 'generates substantial excess heat relative to electrical and chemical inputs'. Readers wishing to make up their own minds should study Dr Tanzella's reports, and listen to a recent podcast (Russell, 2018) in which he speaks about his work. The same site also offers a recent interview with Robert Godes, in which he discusses BEC's latest results.

Some critics will say that Dr Tanzella must be wrong, because the claims are simply so unlikely. That would be an understandable view if BEC's claims were a complete outlier, unrelated to any previous work. But as I said, there's an iceberg's worth of work beneath it, much of it from eminently serious sources (people and institutions). Only someone who hadn't taken the trouble to look at this work could think of BEC as an outlier.

As a very small sample from this iceberg, see McKubre (2009) and Szpak *et al.* (2008) for overviews of long programmes of work by two US laboratories, SRI International themselves and the Space and Naval Warfare (SPAWAR) lab, San Diego, over the 1990s and 2000s; see Takahashi *et al.* (2017) for a short summary of the recent Japanese work mentioned above; and see Beiting (2017) and Mizuno (2017) for two more recent technical papers. All these pieces report results not explicable by known chemical processes. The LENR-CANR.ORG site (LENR-CANR.ORG, 2021) offers many hundreds of other papers.

Finally, for our Norwegian readers, there's a recent 45-page report by the Norwegian Defence Research Institute (Hasvold, 2018). The author, an electrochemist, concludes that in his view, 'LENR is a real phenomenon, the development of which ought to be closely watched'. He says that the alternative 'is to believe in a conspiracy of independent researchers at a number of different institutions', and adds that for the original Fleischmann and Pons reactions in particular, 'the documentation is highly convincing'.

The question I want you to ask yourself — *after* examining some of this material — is not whether you agree with me that BEC has made it over the finishing line specified in my bets. That's an interesting question, but not the important one. The crucial issue is whether LENR in general makes it over a much lower bar, the one that recommends it for serious attention, given how desperate we are for Bill Gates's energy miracle. If you don't agree with me even about the low bar, I'm wondering what you could possibly take yourself to know, that all these authors do not, that could justify such certainty.

If you do agree with me about the low bar, I encourage you to join me in trying to hack the reputation trap. It may be too much to expect mainstream science to scan the horizon very far to port and starboard. That's how science works, and rightly so, in normal circumstances. But if that's where the energy-rich islands might be, that's the direction someone needs to be looking. So, we need some unconventional thinkers — especially young, brilliant, sharp-eyed thinkers — and we need to cheer not sneer at their efforts.

In my view, it is as much a mistake to let reputation blind us to prudence in this case as it was for the icebergs and O-rings. True, it isn't necessarily so catastrophic. But unlike the *Titanic* and the *Challenger*, the planet has all of us on board. So, let's loosen our collars a little, remind ourselves of the virtues of epistemic humility, and do something to encourage our energy mavericks.

For the moment, as cold fusion turns thirty, it remains a black sheep of the scientific family. As the history of science shows us, however, it is often black sheep who bring home the black swans. We don't yet know whether cold fusion will follow the same course, but it is in everyone's interests to show it some warmth. For safety's sake, cold fusion needs to be cool.

3.8 Meeting McKubre

The piece above was published in early March 2019 (Price, 2019). A few weeks later, I met one of the central figures of the field, Dr Michael McKubre. McKubre, a New Zealand-born electrochemist, had worked with Martin Fleischmann in Southampton, some years before the 1989 announcement. He became a leading figure among the small group of scientists who regarded the dismissal of the Fleischmann and Pons results as too hasty. He was Director of the Energy Research Center at SRI International for many years.

By 2019, McKubre was semi-retired, spending most of the year in his home town of Napier, New Zealand. I was on leave in my home town of

Sydney, Australia. I took advantage of our proximity to visit McKubre in Napier. We had dinner together on the thirtieth anniversary of the Pons and Fleischmann announcement in 1989. There were three thirtieth anniversary meetings that month: one at MIT, one at the Russian Academy of Natural Sciences, Moscow, and — most exclusive of all — McKubre and I, with our wives, in Napier, New Zealand.

McKubre gave me a fascinating insider's sense of the history of the field, and of the cold fusion landscape at that point. I was somewhat reassured to hear about the level of work that existed out of the public gaze. I recall particularly that McKubre said, without mentioning any names, that some of the well-known tech companies had quiet programmes.

3.9 Google Joins the Search

This claim was strikingly confirmed a couple of months later, when it became public that Google had been funding LENR work in several universities since 2014. To the surprise of many people, *Nature* published a perspectives piece by some of the Google-funded researchers (Berlinguette *et al.*, 2019). McKubre himself later reported that he had had a hand in initiating that work in 2014, though he had not been involved in it since that time.

This is how the authors of the *Nature* piece present their work:

The 1989 claim of 'cold fusion' was publicly heralded as the future of clean energy generation. However, subsequent failures to reproduce the effect heightened scepticism of this claim in the academic community, and effectively led to the disqualification of the subject from further study. Motivated by the possibility that such judgement might have been premature, we embarked on a multi-institution programme to re-evaluate cold fusion to a high standard of scientific rigour. Here we describe our efforts, which have yet to yield any evidence of such an effect. Nonetheless, a by-product of our investigations has been to provide new insights into highly hydrided metals and low-energy nuclear reactions, and we contend that there remains much interesting science to be done in this underexplored parameter space.

So far, we have found no evidence of anomalous effects claimed by proponents of cold fusion that cannot otherwise be explained prosaically. However, our work illuminates the difficulties of producing the conditions under which cold fusion is hypothesized to exist. This result leaves open the possibility that the debunking of cold fusion in 1989 was perhaps premature because the relevant physical and material conditions had not (and indeed have not yet) been credibly realized and thoroughly investigated. Should the phenomenon happen to be real (itself an open question), there may be good technical reasons why proponents of cold fusion have struggled to detect anomalous effects reliably and reproducibly. Continued scepticism of cold fusion is justified, but we contend that additional investigation of the relevant conditions is required before the phenomenon can be ruled out entirely. (Berlinguette *et al.*, 2019)

Later in the piece, they conclude like this:

Call to action. Fusion stands out as a mechanism with enormous potential to affect how we generate energy. This opportunity has already mobilized a 25 billion dollar international investment to construct ITER [International Thermonuclear Experimental Reactor]. Simultaneous research into alternative forms of fusion, including cold fusion, might present solutions that require shorter timelines or less extensive infrastructure.

A reasonable criticism of our effort may be 'Why pursue cold fusion when it has not been proven to exist?'. One response is that evaluating cold fusion led our programme to study materials and phenomena that we otherwise might not have considered. We set out looking for cold fusion, and instead benefited contemporary research topics in unexpected ways.

A more direct response to this question, and the underlying motivation of our effort, is that our society is in urgent need of a clean energy breakthrough. Finding breakthroughs requires risk taking, and we contend that revisiting cold fusion is a risk worth taking. (Berlinguette *et al.*, 2019)

This message was very congenial indeed from my point of view, of course. But I would add two comments. First, *not revisiting cold fusion* is a risk, too, and potentially a much more serious one. That's the feature that this case shares with the more obvious cases of low-probability high-impact risks — the high potential cost of a false negative.

Second, for the risks that Berlinguette *et al.* (2019) have in mind, the degree of risk depends on sociological factors. Do researchers put their own careers and reputations at risk? If so, we can do something about it, by pushing back against the reputation trap.

The motivation for this Google work had been much the kind of argument I made in my articles. (I'm not trying to claim any credit here. The Google programme pre-dates my first *Aeon* piece, and in any case, I take the point to be obvious to anyone not blinded by the reputation trap.) This was made clear in McKubre's own comment on the Google project, published a few months later.

> Two of the authors of the Perspective article, Ross Koningstein and David Fork, senior engineers at Google, previously wrote an article (Koningstein and Fork, 2014) in which they analyze dispassionately earth's energy situation. In their vision, the known renewable energy sources and any conceivable [source], in their most optimistic projection, cannot supply the energy needs of our planet's growing and advancing population. One of their conclusions is that "new zero-carbon primary energy sources" must be developed. This article appeared in *IEEE Spectrum* in November 2014. Importantly, and before that, rather than congratulating themselves on their analysis and conclusions, the authors set out with Google's support to address that perceived need. The result is what we are discussing today, with the extension enumerated below. Google saw a problem, saw a potential solution, enlisted support and set out to do something about it. (McKubre, 2019, p. 1)

What impact did the revelation of the Google work have on the reputation trap? Some in the LENR community felt that *Nature's* own initial reaction amounted to digging in. The same issue of *Nature* contained both an editorial and a commentary piece concerning the Google work. The latter was by the science writer Philip Ball, himself a former *Nature* editor, and was bylined like this: 'Why revisit long-discredited claims for a source of abundant energy, asks Philip Ball? Because we are still learning how to treat pathological science' (Ball, 2019a, p. 601).

Clearly, that tells *Nature's* readers that cold fusion is pathological science, and is long discredited. (Like 'refuted', 'discredited' is what philosophers call a success word — it takes sides on the facts.) As in the case of the *New Scientist* editorial I mentioned earlier, this byline has the grumpy tone of someone who hasn't yet come to terms with the fact that their earlier hostile judgement may have been too hasty.

Later in the piece, in Ball's own text, the message is qualified, though not by much:

> For some, cold fusion represented a classic example of pathological science. This term was coined in the 1950s to describe a striking claim that conflicts with previous experience, that is based on effects that are difficult to detect and that is defended against criticism by ad hoc excuses. In this view, cold fusion joins an insalubrious list that includes the N-rays of 1903, the polywater affair of the late 1960s and the memory of water episode of the late 1980s.

> *Nature* never published the manuscript by Fleischmann and Pons — the authors withdrew it to focus on follow-up work. But a paper reporting similar findings by a group at Brigham Young University in Provo, Utah, was published in April of that year (Jones *et al.*, 1989). The only report at the time from Fleischmann and Pons was a short paper, lacking in detail, in the *Journal of Electroanalytical Chemistry* (Fleischmann and Pons, 1989).

> *Nature* did publish follow-up studies by other groups, including one that used the actual equipment of Fleischmann and Pons (Salamon *et al.*, 1990). None observed any hint of cold fusion, and no convincing evidence has since materialized. (Ball, 2019a, p. 601)

Friends of LENR might respond that it is hardly surprising that none of the pieces published in *Nature* reported any hint of cold fusion, given the stance that *Nature* soon took on the subject. It was in the light of this well-known stance that the *Nature* Perspective piece in 2019 came as such a surprise. An effect of the stance had long been that positive results had to go elsewhere — to a few journals ready to take the reputational risk, a risk that *Nature* itself had done much to create.

Melinda Baldwin's excellent recent history of *Nature* gives this account of the episode:

> Although Pons and Fleischmann had indeed submitted an article to *Nature*, that journal never printed it: only Steven Jones's article, with its far more modest claims about neutron production and excess heat from the reaction, would be published in *Nature*. Instead of being the forum where a new era of energy was declared, *Nature* quickly became a major

center of cold fusion skepticism. By 29 March 1990, a year to the week after the first mention of cold fusion in *Nature*, [the Editor John Maddox] felt secure enough to declare "Farewell (Not Fond) to Cold Fusion" in the magazine's leader. (Baldwin, 2015, p. 201)

As Baldwin goes on to say:

During the cold fusion controversy, Maddox … and the rest of the editorial staff cast the cold fusion episode as a battle between careful, peer-reviewed, properly conducted science and sloppy science revealed through press conferences in hopes of wealth through patents. Maddox wrote editorials criticizing Pons and Fleischmann's methods, associate editor David Lindley wrote news articles forecasting the death of cold fusion, and the journal's editorial staff gave significant space to cold fusion's most prominent scientific critics. Where *Nature* led, science reporters followed. News outlets such as *Time*, the *Economist*, and the *Wall Street Journal* all covered *Nature's* role in the cold fusion controversy and portrayed the journal's skepticism as proof that the scientific community was rejecting the Pons-Fleischmann claims. Ultimately, the cold fusion episode convinced many observers of the scientific journal's continued importance to the scientific community and illustrated *Nature's* influence among both scientists and laymen at the end of the twentieth century. (Baldwin, 2015, pp. 201–202)

Returning to Ball's commentary, the claim that 'no convincing evidence has since materialized' is clearly a judgement call, with which McKubre amongst others would simply disagree. Again, Ball chooses words that seem to leave no room for doubt on the matter.

A more important failing, in my view, is that nothing in this commentary acknowledges the issues of epistemic risk, to which the Google team are clearly sensitive — the potentially catastrophic cost of a false negative, in a case like cold fusion. *Nature* is one of the few institutions that could, if it chose, popularise this view, and hence do something about the reputation trap. Judged by this standard, I felt that Ball's commentary was a missed opportunity.

The editorial in the same issue does a little better. In this case, the byline does not immediately close the door: 'Major project to reproduce controversial claims of bench-top nuclear fusion kindles debate about when high-risk research is worthwhile', though again, it misidentifies the important risk, which is that of *not* doing the research, given the high potential cost of a false negative. But the piece does go on to note that '[s]ociety's need for cheaper and cleaner sources of energy is more pressing than ever, and, if cold fusion were possible, it could be a disruptive technology with a world-changing pay-off' (Anonymous, 2019a).

The editorial concludes like this:

> The [Google] team found no evidence whatsoever of cold fusion.
>
> Is that the final nail in the cold-fusion coffin? Not quite. The group was unable to attain the material conditions speculated to be most conducive to cold fusion. Indeed, it seems extremely difficult to do so using current experimental set-ups — although the team hasn't excluded such a possibility. So the fusion trail, although cooling, is not yet cold, leaving a few straws for optimists to clutch on to.
>
> The question now is whether it is even worth continuing this research. Here, the message is more nuanced. The project has produced materials, tools and insights — such as calorimeters that operate reliably under extreme conditions, and techniques for producing and characterizing highly hydrided metals — that could benefit other areas of energy and fusion research. But whether the spin-off benefits alone justify continued efforts and investment in pursuit of a probable pipe dream is another matter. Opinions are split.
>
> So what do we take home from a multi-year failed experiment? First, that the programme has been conducted with rigour and attention to detail — we can have confidence in the results. Second, although the work provides no support for fringe groups that continue to insist that cold fusion exists, it does bring this research area back into the light of harsh scientific scrutiny. And, by doing so, the project might help responsible research in this general area to become less taboo, even if the chances of achieving cold fusion still look extremely remote. (Anonymous, 2019a)

This doesn't miss the opportunity to wheel out some of the tropes of the reputation trap, and certainly ignores contrary evidence even from neutral expert assessors (more on this in Section 9.5). Some might feel that the remark about helping 'responsible research in this general area to become less taboo' is a little rich, coming from the journal that did so much to foster that view in the first place. But despite all that, the editorial does concede that the door is ajar.

A few months later, after meeting some of the Google team's researchers, Philip Ball published a second piece, this time in a journal with less of a horse in the race than *Nature* (Ball, 2019b).[4]

> [C]old fusion has never gone away. A few researchers, working at the fringes of the scientific community, have continued to claim to see tantalizing signs that there really is something in it after all. However, the field has never shaken off its bad reputation. There was much surprise when in June, 30 years after the original event, *Nature* published an article by a team of researchers funded by Google describing renewed searches for "low-energy" fusion of hydrogen isotopes (deuterium, which has a lower energy threshold for fusion than hydrogen-1) using palladium electrodes.
>
> The paper reported no evidence of such a process in electrochemical experiments similar to those of Pons and Fleischmann, but it described a low level of fusion from a different experimental setup in which a plasma of deuterium ions surrounded a negatively charged palladium wire. The new findings will not persuade anyone that Pons and Fleischmann were right, but they could give cold fusion a new lease on life. Moreover, the study showed that there are interesting things still to learn about the materials science of the palladium–hydrogen system. (Ball, 2019b, p. 883)

Most interestingly, Ball also spoke to some of the scientists involved in the Google-funded work. Here, he quotes Curtis Berlinguette, head of the Berlinguette research group at UBC, Vancouver, and lead author on the *Nature* Perspectives piece:

> "Renewable energy and fusion technologies are not scaling at the pace we need them to," says Berlinguette. "If cold fusion were realizable, it could take the world into an era of energy surplus rather than scarcity. It therefore seemed irresponsible to not take another look at it. For me,

---

[4]At around the same time, *Nature Materials* published an editorial entitled 'Coming in from the Cold', that is also clearly informed by a good understanding of the Google team's work (Anonymous 2019b).

cold fusion started in 2015," says Berlinguette. "Prior to that, I didn't know enough about it to have an opinion. I was driven simply by curiosity to learn more about the field." (Ball, 2019b, p. 883)

There is much to like here, from my point of view, especially Berlinguette's ability to look beyond the reputational factors to the true imperatives of the case ('It ... seemed irresponsible not to take another look at it'). His epistemic attitude is also admirable. My experience in this field is that 'not knowing enough about it' doesn't prevent people from having very strong opinions — hence my plea for 'epistemic humility' at the end of my 2019 piece.

Berlinguette's webpage now says this: 'His program also likes to work on high risk, high impact clean energy projects like cold fusion' (UBC, 2021). This is huge progress, in my view, in the sociological sense. In particular, it provides cover for younger scientists — the 'young, brilliant, sharp-eyed thinkers' I had in mind in Section 3.7* (Price, 2019) — to work on these issues, without such a risk to their own careers.

Has Google's own reputation provided cover, in this sense, for other research groups in the field? Not surprisingly, the answer is a resounding 'yes'. With this in mind, let's turn to other developments in the field since 2019.

3.10 The Global LENR Landscape — Updates Since 2019

3.10.1 *Japan*

Let's begin with Japan. The Clean Planet initiative, first mentioned in my postscript from 2017 (Section 3.5*), continues to give the impression of steady progress, in two dimensions: both technological progress and striking reputational progress, evident in Clean Planet's standing with significant organisations outside the LENR field.

Taking the latter dimension first, Clean Planet now has equity participation of two major Japanese companies. One is Mitsubishi Estate, a member of the Mitsubishi group. The second is Miura Co. Ltd, a major Japanese boiler manufacturer.

Miura invested in Clean Planet in May 2019. More recently, in September 2021, it signed an agreement with Clean Planet 'to jointly develop industrial boilers' (Miura, 2021) based on Clean Planet's technology. This description of Clean Planet is from Miura's own press release about the collaboration. Note the reference to 'a major US IT company':

> Clean Planet Inc. is a venture enterprise that has worked on the research and development of Quantum Hydrogen Energy, [... which] is currently attracting attention globally, and large companies and investors representing every industry are beginning full-scale participation in this field, as can be seen from the entry of a major US IT company. Against this background, Clean Planet Inc. has embarked on a range of cutting-edge research and development efforts in collaboration with Tohoku University, and in April 2021 began work on development for practical use for release of one kilowatt of thermal energy using Quantum Hydrogen Energy. (Miura, 2021)

This news has received some coverage in the mainstream Japanese business press, though not as far as I know in the West. This is from the NikkeiBP website:

> Focusing on [Clean Planet's] research results, Mitsubishi Estate invested in Clean Planet in January 2019 and Miura Co., Ltd. invested in Clean Planet in May of the same year. Since then, research has progressed steadily toward practical use, so [Clean Planet] decided to start full-scale joint development with Miura Co., Ltd. regarding its application to industrial boilers. A prototype will be produced in 2022 and will be commercialized in 2023. (NikkeiBP, 2021)

An excellent impression of Clean Planet's apparent progress over the past decade is conveyed by Figure 1, reproduced with permission from the Clean Planet website (Clean Planet, 2021). Note, in particular, the reported timings of completion of a 100-W 'model device' (2018) and 'starting testing' of 1-kW prototypes (May 2021).

I would like to encourage readers to consider the possible explanations for the existence of this diagram, and the claims it embodies. So far as I can see, there are essentially three possibilities:

(i) We take the claims at face value, and accept that Clean Planet does have devices producing large amounts of heat, not explicable in chemical or other nonnuclear terms. This means accepting that

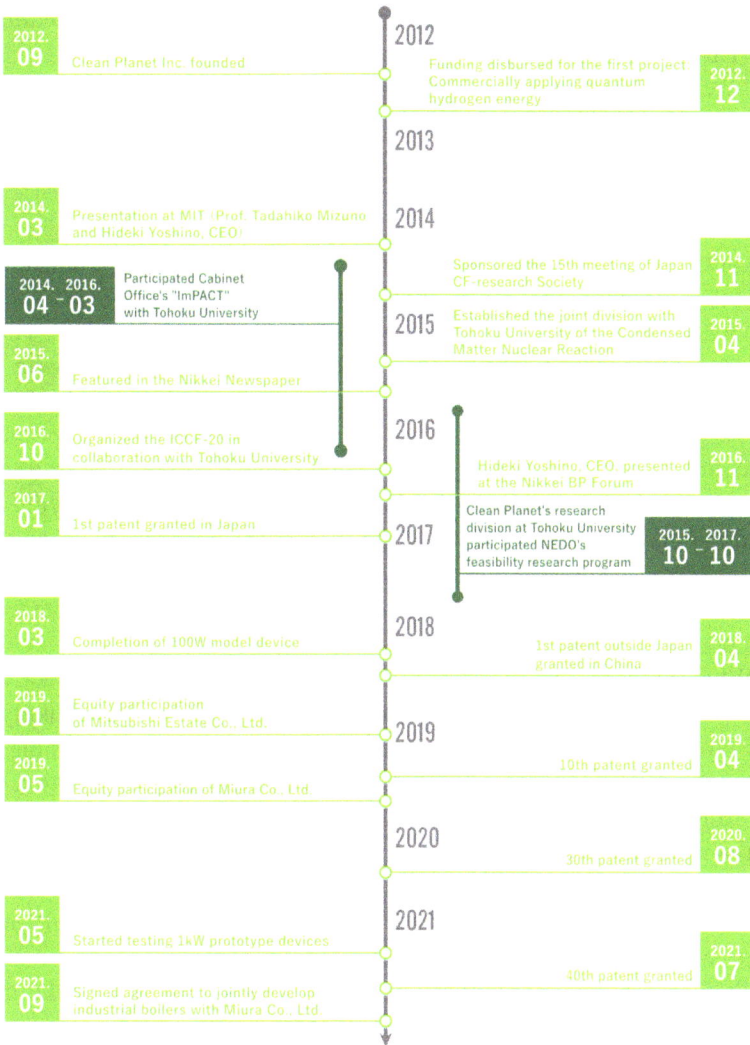

Figure 1.   Clean Planet timeline (from Clean Planet, 2021).

so-called cold fusion is real (even if perhaps badly named — this semantic point would need to await a good theoretical explanation of the excess heat in question).

(ii)   We accept that the claims are sincerely made, but take them to result from gross error of some kind — either measurement error or failure to spot an alternative explanation for the results, not requiring LENR.

(iii) We deny that the claims are sincerely made, but take them to result from deception or misrepresentation of some kind (either by Clean Planet personnel themselves, or by someone else — perhaps the entire website is a hoax by high school hackers, for example).

Unless there is some further possibility that I have missed, it is a rational requirement that you divide your degrees of credence between the three possibilities (i)–(iii) in a way that adds up to 1 or 100%. (If you don't do that, a clever Dutch bookie can make a combination of bets with you that you are guaranteed to lose, no matter what — see Vineberg, 2016.) I encourage readers who continue to give negligible credence to option (i), the reality of cold fusion, to consider what credence they wish to give to the other two possibilities. Please don't dodge the question by saying that you don't have access to the evidence. Take the diagram itself as your evidence. You do have access to the diagram — that's why I have presented the options in this way. Ask yourself what credence you give to the three possible explanations for *its* existence.

What do I think? In my view option (iii) is vanishingly unlikely, in this case. Gross error is possible, but in my view now rather unlikely — even more so than in 2019, given the reported progress concerning the size of the effect. So, I now give less than 20% credence to option (ii). Accordingly, I give at least 80% credence to the option (i), that of taking the claims at face value.

As I said with respect to Brillouin in Section 3.7* (Price, 2019), resistance to this conclusion would be understandable if Clean Planet's claims were complete outliers, unrelated to any previous scientific work. But they're not. Like Brillouin, Clean Planet is simply one of the most prominent tips of a substantial iceberg of scientific work, over thirty years.

Readers may also feel sceptical on the grounds that if there were something to Clean Planet's claims, we would have heard a lot more about them. Where was the news of Clean Planet's collaboration with Miura in the month of COP26 in Glasgow, for example? But you know the answer to this. The reputation trap continues to ensure that LENR remains unmentionable by most journalists. Would you put your head above the parapet and be the first major science writer to be seen to be taking it seriously?

3.10.2 *Europe*

Turning now to Europe, I know of no recent scientific or technological progress to compare with that in Japan. But the sociological dimension is a different matter. The EU's flagship Horizon 2020 funding programme recently awarded two large grants to multi-institution teams working on LENR. This is a very significant shift, in my view.

The first of these teams, the HERMES project, has been funded at €3,999,870 over five years from 2020–2024. Coordinated from Finland, HERMES is described like this on the funder's website (the leverage provided by the Google work is again explicit):

> In 1989, electrochemists Martin Fleischmann and Stanley Pons made headlines with their claim to have produced excess heat using a simple apparatus working at room temperature. Their experiment involved loading deuterium in a palladium metal. As many experimenters failed to replicate their work, cold fusion remains a controversial topic in the scientific community. Nevertheless, a vociferous minority still believes in this elusive phenomenon. Since 2015, Google has been funding experiments into cold fusion. Although no evidence has been found for this phenomenon, it is clear that much pioneering research remains to be conducted in this poorly explored field. The EU-funded HERMES project will employ advanced techniques and tools developed over the last few decades to investigate anomalous effects of deuterium-loaded palladium at room and intermediate temperatures. (CORDIS, 2021a)

The second EU grant, from the same Horizon 2020 funding programme, is for the CleanHME project. This was funded at €5,678,597.50 over 2020–2024. It is coordinated from Poland and described like this:

> With climate change being a major global concern in recent times, new efficient clean energy sources are in high demand, and there has been a rise in the use of many of them, such as solar or wind generators. One very promising energy source is hydrogen–metal energy (HME), which could be used for small mobile systems as well as in stand-alone heat and electricity generators. Unfortunately, little research has been conducted concerning HME. The EU-funded CleanHME project aims to change this. It will produce an elaborate comprehensive theory of HME

phenomena that would assist in the optimisation of the process and construct a compact reactor to test HME technology. (CORDIS, 2021b)

Again, these grants indicate a very welcome change in the sociological climate, in my view.

3.10.3 *USA — New results from a longstanding NASA programme*

Moving now to the US, our first stop is at NASA. There has been some interest in LENR within NASA since at least the early 2010s. For example, Wells *et al.* (2014) reports on a NASA-funded project investigating the potential of LENR as a power source for terrestrial flight, while Bushnell *et al.* (2021) is a NASA report on its potential for space flight. But a much more hands-on approach to the subject came to light in 2020, as a NASA website reports:

NASA Detects Lattice Confinement Fusion. A team of NASA researchers seeking a new energy source for deep-space exploration missions, recently revealed a method for triggering nuclear fusion in the space between the atoms of a metal solid. Their research was published in two peer-reviewed papers in the top journal in the field, *Physical Review C* (Pines *et al.*, 2020; Steinetz *et al.*, 2020) …

Nuclear fusion is a process that produces energy when two nuclei join to form a heavier nucleus. "Scientists are interested in fusion, because it could generate enormous amounts of energy without creating long-lasting radioactive byproducts", said Theresa Benyo, Ph.D., of NASA's Glenn Research Center. "However, conventional fusion reactions are difficult to achieve and sustain because they rely on temperatures so extreme to overcome the strong electrostatic repulsion between positively charged nuclei that the process has been impractical."

Called Lattice Confinement Fusion, the method NASA revealed accomplishes fusion reactions with the fuel (deuterium, a widely available non-radioactive hydrogen isotope composed of a proton, neutron, and electron, and denoted "D") confined in the space between the atoms of a metal solid. In previous fusion research such as inertial confinement fusion, fuel (such as deuterium/tritium) is compressed to extremely high levels but for only a short, nano-second period of time, when fusion can occur. In magnetic confinement fusion, the fuel is heated in a plasma to temperatures much higher than those at the center of the Sun. In the new method, conditions sufficient for fusion are created in the confines of the metal lattice that is held at ambient temperature. While the metal lattice, loaded with deuterium fuel, may initially appear to be at room temperature, the new method creates an energetic environment inside the lattice where individual atoms achieve equivalent fusion-level kinetic energies. (NASA Glenn Research Center, 2021)

We learn from this that the NASA team are as skilled at splitting hairs as they are at physics. Koziol (2020) picks up this aspect of the story at *IEEE Spectrum:*

"What we did was *not* cold fusion," says Lawrence Forsley, a senior lead experimental physicist for the project. Cold fusion, the idea that fusion can occur […] in room-temperature materials, is viewed with skepticism by the vast majority of physicists. Forsley stresses this is hot fusion, but "We've come up with a new way of driving it."

"Lattice confinement fusion initially has lower temperatures and pressures" than something like a tokamak, says [Theresa Benyo, an analytical physicist and nuclear diagnostics lead on the project]. But "where the actual deuteron-deuteron fusion takes place is in these very hot, energetic locations." (Koziol, 2020)

It is not clear to me if the notion of 'locally hot' makes sense at this scale. Temperature is a statistical property, defined in terms of the mean kinetic energy of the atoms or molecules of a substance. For individual molecules, atoms, or subatomic particles, it is more appropriate to speak of their energy than their temperature.

In any case, these remarks are clearly an attempt by the NASA group to protect themselves from the heat of the reputation trap. Forsley himself has long been a significant member of the cold fusion community, appearing with other major figures in the field as co-author (Marwan *et al.,* 2010), for example. There is an excellent interview with him, discussing this recent NASA work at Hughes (2020).

One imagines the ghost of Martin Fleischmann rolling his eyes, and pointing out that what the NASA group have in mind is perfectly compatible with his own use of the term cold fusion, which never referred to anything more than the temperature of the test tube — and then being ticked off by sceptics, keen to cover their own retreat, for 'not making that clear in the first place!'[5]

3.10.4 *New naval manoeuvres*

Turning from deep space to the deep sea, the US Navy has long supported research on cold fusion. In Section 3.7* (Price, 2019), I mentioned work by the Space and Naval Warfare (SPAWAR) lab in San Diego. The combination of the NASA news and the Google news apparently inspired a different Navy group to pull together a collaboration of US Government labs, as another *IEEE Spectrum* piece describes. Koziol (2021) describes how the Indian Head Division of the Naval Surface Warfare Center has brought together a team of Navy, Army, and National Institute of Standards and Technology, to 'try and settle' the LENR debate.

Koziol notes that the project was prompted by the Google interest in the area, and explains how the labs involved felt about the controversial nature of the topic:

> "We got our impetus from the Google paper appearing in *Nature*," says Carl Gotzmer, Indian Head's Chief Scientist. [...] [T]he Indian Head team decided that, as a government lab, they had a little more freedom to pursue a controversial topic, so long as it also offered up the prospect of rewarding scientific results. (Koziol, 2021)

He quotes Oliver Barham, an Indian Head project manager, on the project's attitude to the controversy:

> "I'm not as worried about looking into something that is considered controversial as long as there's good science there," says Oliver Barham, a project manager at Indian Head involved in the effort. "The whole point of our effort is we want to be doing good science." (Koziol, 2021)

---

[5]I believe that Fleischmann and Pons did not use the term 'cold fusion' themselves, initially, though it was soon used to describe their claims.

3.10.5 *The ARPA-E workshop* (*October 2021*)

One of the most significant actors in the story of cold fusion is the US Department of Energy (DOE). The DOE produced reports on the field in 1989 and 2004. The Advanced Research Projects Agency-Energy (ARPA-E) is the DOE's version of the well-known Defense Advanced Research Projects Agency (DARPA) — the agency 'credited with such innovations as GPS, the stealth fighter, and computer networking', as the ARPA-E website puts it. ARPA-E itself says that it 'advances high-potential, high-impact energy technologies that are too early for private-sector investment' (ARPA-E, 2021a).

In October 2021, ARPA-E hosted a publicly announced workshop on LENR. This is how it is described on its own website:

> The objective of this workshop was to explore compelling R&D opportunities in Low-Energy Nuclear Reactions (LENR), in support of developing metrics for a potential ARPA-E R&D program in LENR. Despite a large body of empirical evidence for LENR that has been reported internationally over the past 30+ years in both published and unpublished materials, as well as multiple books, there still does not exist a widely accepted, on-demand, repeatable LENR experiment nor a sound theoretical basis. This has led to a stalemate where adequate funding is not accessible to establish irrefutable evidence and understanding of LENR, and lack of the latter precludes the field from accessing adequate funding. Building on and leveraging the most promising recent developments in LENR research, ARPA-E envisions a potential two-phase approach toward breaking this stalemate: (1) Support targeted R&D toward establishing at least one on-demand, repeatable LENR experiment with diagnostic evidence that is convincing to the wider scientific community (focus of this workshop); (2) If phase 1 above is successful (metrics to be determined), support a broader range of R&D activities (to be defined later) toward better understanding of LENR and its potential for scale-up toward disruptive energy applications, thus setting up LENR for broader and more systematic support by both the public and private sectors. (ARPA-E, 2021b)

The website adds that for the purposes of the workshop, 'LENR is defined as a not-yet-understood process (or class of processes) characterized by system energy outputs characteristic of nuclear physics (typically >> 1 keV/amu/reaction) and energy inputs characteristic of chemistry (~eV/atom)'.

The website now includes links to workshop presentations from a number of speakers, including several who will be familiar to readers who have made it this far: Michael McKubre, the NASA team, Clean Planet, and Brillouin. While it is too soon to know what the impact of this workshop will be, it does seem to represent a further welcome shift in the sociological climate. Concerning the physics, I particularly recommend the presentation by Florian Metzler (MIT), which does a very interesting job of pulling together various strands in the LENR literature (Metzler, 2021).

There is also a presentation from ARPA-E Fellow, Dr Katherine Greco, about the reviews of the field conducted by DOE in 1989 and 2004. Although the LENR community was disappointed by the 2004 review, it was certainly not wholly negative. As Greco's presentation puts it, the DOE panel was 'nearly unanimous' that

> funding agencies should entertain individual, well designed-proposals [into]
> – Whether or not there is anomalous energy production in Pd/D systems
> – Whether or not D-D fusion reactions occur at energies ~eV. (Greco, 2021)

Some LENR researchers complain that this recommendation was not actually followed — that funding proved as elusive as before (Maguire, 2014). For my purpose, however, what's relevant is that the DOE's panel of 18 rather sceptical physicists did not simply dismiss the field. Their individual views varied. McKubre (Maguire, 2014) reports that the nine who participated in an in-person one-day meeting were almost uniformly positive. The report itself states that on the question of 'experimental evidence for the occurrences of nuclear reactions in condensed matter at low energies (less than a few electron volts)', one third of the reviewers were either completely convinced or 'somewhat convinced' (DOE, 2004). And collectively, in any case, they left the door ajar. In contrast to *Nature* much more recently, the DOE reviewers did *not* treat LENR as 'long-discredited … pathological science' (Ball, 2019a).

Meanwhile, the DOE itself has just announced plans to establish a new Office of Clean Energy Demonstrations:

> President Biden's Bipartisan Infrastructure Law provides more than $20 billion to establish the Office of Clean Energy Demonstrations and support clean energy technology demonstration projects in areas including clean hydrogen, carbon capture, grid-scale energy storage, small modular reactors, and more. Demonstration projects prove the effectiveness of innovative technologies in real-world conditions at scale in order to pave the way towards widespread adoption and deployment. The founding of this office represents a new chapter that builds on DOE's long-standing position as the premier international driver for clean energy research and development, expanding DOE's scope to fill a critical innovation gap on the path to net-zero emissions by 2050. (DOE, 2021)

In light of the DOE's reluctance to fund LENR research in 1989 and (as it has turned out) since 2004, it will be interesting to see whether it is now actually moving in a different direction. Certainly, the emphasis here on 'demonstrations' chimes very nicely with what the ARPA-E workshop had in mind.

3.10.6   *The US Defence Intelligence Agency (2009)*

To balance this impression of *new* interest, let me end this list of developments involving US Government agencies with newly available evidence of much earlier interest. In response to a freedom of information request, the US Defence Intelligence Agency has recently released a 2009 report titled 'Technology Forecast: Worldwide Research on Low-Energy Nuclear Reactions Increasing and Gaining Acceptance' (DIA, 2009).

This report includes a detailed list of work undertaken in the period 1989–2009, in many parts of the world. It then concludes like this:

> Although no one theory currently exists to explain all the observed LENR phenomena, some scientists now believe these nuclear reactions may be small-scale deuterium fusion occurring in a palladium metal lattice. Some others still believe the heat evolution can be explained by non-nuclear means. Another possibility is that LENR may involve an

intricate combination of fusion and fission triggered by unique chemical and physical configurations on a nanoscale level. **This body of research has produced evidence that nuclear reactions may be occurring under conditions not previously believed possible.** (DIA, 2009, emphasis in the original)

The report goes on to discuss potential applications of LENR, 'if nuclear reactions in LENR experiments are real and controllable'. When I read it, I recalled the scientists I mentioned in my original *Aeon* piece, who were technology forecasters for a similar UK organisation, and who told me that they had trouble getting their agency to take LENR seriously, even as a low-probability possibility. It turns out the work they needed had been done by some of their US counterparts, several years earlier.

3.11 Summary — The State of the Field

In my original *Aeon* piece (Price, 2015), I expressed this view:

> I suspect it's too late to dismantle the [reputation] trap for LENR – the horse is already in the process of bolting, I think. If so, then the field is going to be mainstream soon, in any case. But we could try to learn from our mistakes. There may be other potential cases with a similar payoff structure (a high cost for false negatives, with a low cost for false positives).

This now seems to me to have been mistaken in one respect: I chose the wrong creature, or at least the wrong gait. In light of the rate of progress I hoped for then, LENR has been more of a stroll than a gallop. Still, we know what slow and steady does, and this is still what I expect for LENR. That is, I expect that the science of LENR (defined, let us say, in the terms specified by the recent ARPA-E workshop) will indeed become mainstream. It is too soon to tell whether it will turn out to be useful as a source of energy, but I think the present signs are quite encouraging about that, too.

As I have noted at several points, it is not unusual in the history of science for formerly controversial ideas to become mainstream. So, if LENR goes the way I expect, it will not be exceptional on those grounds alone. If there is a case for regarding it as exceptional, in the long run, I think it will rest on two main factors.

The first factor will be the depth of the reputational trap from which it will have managed to dig itself out. Here, I'm thinking of several things: the severity of the condemnation and ridicule to which the field has been subject, its highly public nature, and the involvement of major scientific institutions such as *Nature* in administering it. I stress that I'm not making a critical judgement at this point about people or institutions that played a part in giving the field this reputation; I'm aware, of course, that there's a case to be made that Fleischmann and Pons set themselves up for it, in choosing their own public and unconventional course.

Where I do make a critical judgement concerns the second factor that will make LENR exceptional, if it does return to the mainstream. This is the point I have stressed from the beginning, about the high cost of a false negative. I mentioned Heather Douglas, who spoke at our maverick workshop at Trinity College in 2017. This is from a piece by Douglas called 'Rejecting the Ideal of Value-Free Science' (Douglas, 2007):

> In general, if there is widely recognized uncertainty and thus a significant chance of error, we hold people responsible for considering the consequences of error as part of their decision-making process. Although the error rates may be the same in two contexts, if the consequences of error are serious in one case and trivial in the other, we expect decisions to be different. Thus the emergency room avoids as much as possible any false negatives with respect to potential heart attack victims, accepting a very high rate of false positives in the process. ... In contrast, the justice system attempts to avoid false positives, accepting some rate of false negatives in the process. Even in less institutional settings, we expect people to consider the consequences of error, hence the existence of reckless endangerment or reckless driving charges.

Douglas goes on to discuss the possibility that '[w]e might decide to isolate scientists from having to think about the consequences of their errors', but rejects it. She argues that 'we want to hold scientists to the same standards as everyone else', and therefore 'that scientists should think about the potential consequences of error' (Douglas, 2007).

In my view, the blameworthy feature of the cold fusion case, if it becomes mainstream — indeed, I think it is blameworthy, *whether or not* it becomes mainstream — is the apparent failure of many of its critics to take this 'Douglas Doctrine' into account. Of course, it is too soon to judge whether this has made any practical difference. We don't yet know whether LENR will turn out to be a useful energy source. Even if so, it will be difficult to estimate how long a delay the treatment of the field might have caused.

But these unknowns are in one sense irrelevant. We don't excuse lax safety practices just because a disaster fails to happen as a result. Think of Douglas's examples — reckless endangerment and reckless driving. A person may be guilty of these things, even if by good fortune they fail to harm anyone. I think there's a *prima facie* case for charging some of the institutions of science with reckless endangerment, or something on that spectrum, in the case of LENR.

However, it would take a much better historian than me to write a detailed brief for such a charge, or indeed to try to make the case for the defence. Did actors such as John Maddox at *Nature* ever give consideration to the costs of a false negative, for example? Rather than trying to answer that historical question, let me turn to a more useful one. How can we do better in future cases with a similar risk profile? I have several suggestions.

3.12 Recommendations

As I said at the beginning, much of my interest in the case of cold fusion stems from its similarities with other cases in which hasty dismissal of an unconventional scientific claim might be dangerous. In particular, I have in mind the cases in which the claim in question concerns a potential catastrophic risk. My recommendations here are offered with all of these cases in mind.

3.12.1 *Foregrounding the Douglas Doctrine*

The most obvious recommendation is that the factors to which Heather Douglas calls attention need to be better known. They need to be internalised both by scientists themselves and by important scientific institutions, such as major journals and learned societies. As Douglas's examples show, the points themselves are not difficult to see, and are already built into standard practice in many contexts, such as emergency rooms and law courts. But they need to be much more familiar, and to be encapsulated in simple maxims that come to seem a matter of common sense.

By way of comparison, think of the principle that correlation is not causation. That rolls so easily off the tongue, these days, that no one has any excuse for ignoring it. Imagine the incredulous response to someone who does so: 'You didn't *realise* that correlation need not imply causation? Where have you been?'

If we had one or two similarly familiar phrases capturing the principle that tolerance for error needs to depend on what's at stake, it would be harder for anyone to ignore them. I've called it the Douglas Doctrine in the hope that in this case, too, some catchy alliteration will help. But a catchy descriptive version would be even better. In Section 3.7* (Price, 2019), I put it this way: 'The more disastrous a potential failure, the more improbable it needs to be before we can safely ignore it', Perhaps 'High stakes/low bar' might serve as an easily memorable version of this principle?

3.12.2 *Understanding reputation traps*

My second recommendation is that scientists and scientific institutions need a better understanding of the way in which reputation operates in science, especially negative reputation. Hopefully, such an understanding would provide a degree of self-awareness, a willingness to consider whether one's own reactions are to be trusted. The pathology of the reputation trap, as I've called it, is that reputational factors get in the way of listening to the 'low-probability voices', to the mavericks who need to be heard when a lot is at stake. If this were well understood, it would be much easier to take steps to avoid it.

Obviously, this recommendation combines with the first. The high-stakes/low-bar principle might tell us that a fringe view needs to be taken seriously, while an understanding of the grip of the reputation trap might make that easier.

3.12.3 *Improving the climate of scientific debate*

Understanding the pathology of reputation traps is one thing, but my third recommendation is that we do something to discourage them in the first place. I think we need to pay attention to the language and 'climate' of scientific debate. In particular, we need to be conscious of the role of what we can call *epistemic slurs*. I use this term in the sense of the philosopher

Josh Habgood-Coote, who argues that we should 'stop talking about "fake news"' (Habgood-Coote, 2018, 2019):

> "Fake news" has a rich expressive meaning, and often functions as an *epistemic slur.* Applying "fake news" to a news story seems not to describe the story, but to express disdain toward the story, the institution that produced it, and (in some cases) toward people that believe the story.

In my case, I have in mind terms such as 'crank', 'crackpot', 'chicanery', 'pseudoscience', 'conspiracy theory', and 'debunk'. These terms do have some descriptive content, to varying degrees, but all of them also function as insults. They 'express disdain' towards a person or a view, as Habgood-Coote puts it. In other words, the target is accused not merely of an *epistemic* deficit — for example, of making claims that are insufficiently supported by evidence — but of what we might call a connative deficit, as well. 'He's not just mistaken, he's a crank!'

When used effectively, these terms thus function to put their targets in categories defined by our disdain towards their members. It is worth asking whether this is necessary, or helpful, in scientific debates. After all, one obvious consequence of the availability of such labels is that they are easily used by others, including many who are not competent to participate in the scientific debate themselves.

A possible remedy would be to attempt to flag and deprecate the use of such epistemic slurs — to make them seem unacceptable, or at least less acceptable, in scientific argument. We already do this with slurs of other kinds, in other contexts — think of prohibitions on so-called hate-speech. Many workplaces have guidelines in place to discourage verbal bullying (as well as other kinds of bullying).

In my own discipline, there has been a recent movement to improve the culture of discussion in Q&A after research talks and similar events; see NYU Philosophy (2021), for example, for the kind of guidelines that are now common. This has been a response to the complaint that many people found some aspects of the old culture hostile and intimidating. Something similar could be done in science, I think. We would end up presenting a less hostile face to views with which we disagreed, because some of the ways of signalling hostility had been discouraged.

It might be objected that such a move would make it harder to police the boundaries of science, to separate good science from bad. But if so, in my view, it is not clear if that would be a bad thing. Think of slurs as a form of verbal violence. Real police officers no doubt find it easier to police boundaries (e.g., to separate protestors from members of the public) if they are allowed to resort to violence, but that doesn't mean it is a good thing. Separating good science from bad certainly matters, but that doesn't mean that 'anything goes' in the attempt to do so.

Like physical violence, slurs also make it easier for people and institutions to erect barriers in places that suit their own (non-epistemic) interests. If cold fusion had turned out to be 'the real deal' in 1989, many other research programmes and economic activities would have been massively affected. Did such considerations ever get in the way of a fair hearing for cold fusion? That's another historical question I'll leave to one side, but it would hardly be surprising if factors like these had played a role. Think of the (now) well-known opposition to climate science over recent decades, e.g., by fossil fuel companies.

My point here is a different one. Epistemic slurs provide a very powerful weapon for actors who are not motivated simply by getting the science right, enabling them to attack the *reputation* rather than the *arguments* of their opponents. This is a version of the famous *ad hominem* fallacy (Wikipedia Contributors, 2021d), of course. Like that fallacy in other contexts, it is often a sadly effective rhetorical device. In my view, it would be very much in the interests of good science to lessen this risk, by shining a disapproving light on the language that facilitates it.

Clearly, this point is very general. Many commentators have noted the degeneration in the climate of public scientific debate, especially in the COVID-19 epidemic. If I have a novel claim, it is only a reminder that these issues take on a particular urgency in cases in which the costs of wrongful dismissal are exceptionally high. In other words, one corollary of the Douglas Doctrine is that we need to be especially careful of epistemic slurs in cases with this value profile.

3.12.4 *Epistemic humility*

At the end of Section 3.7* (Price, 2019), I recommended that we should 'loosen our collars a little, remind ourselves of the virtues of epistemic humility, and do something to encourage our energy mavericks'. What did I have in mind by the phrase 'epistemic humility'? A couple of things, one general and one particular.

The general point is that if we want to pay attention to low-probability options, then we need to leave the door ajar. We can't afford to take mainstream confidence that those views are mistaken as a reason for excluding them from the scientific conversation. Accordingly, an appreciation that one's own strongly held beliefs might be mistaken is an important virtue in this kind of context. This is the main thing I had in mind by epistemic humility; see Angner (2020) for a similar plea for epistemic humility in the COVID-19 pandemic.

The particular point I had in mind was a view about some of my friends and discussants, who are much more sceptical than me about cold fusion. I felt that they were insufficiently attentive to the evidence provided by better-placed observers, and I regarded this as a failure of epistemic humility. In Price (2016), I expressed the thought like this:

> My recommendation to these friends is to keep at least one eye open. For herd animals – like ostriches and scientists – a good way to know when to move is to keep an eye on peers who are closer to the action. If they start to shift, then you should consider it too – unless you have good reason to think that you know something that they don't.

I returned to this kind of thought at the end of Section 3.7* (Price, 2019). I list a number of publications by (apparently) well-qualified authors, all of whom clearly believe that LENR should be taken seriously, and that they themselves have the evidence to support such a view. I ask the reader whether they feel that this material lifts the subject above the very-low-probability bar that would recommend LENR for serious attention, in the light of the high-impact/low-bar principle. I say, 'If you don't agree with me even about the low bar, I'm wondering what you could possibly take yourself to know, that all these authors do not, that could justify such certainty?'

In Section 3.6, I suggest setting the bar at 5%, which I felt was generous to my opponents. (A probability of 5% would justify an investigation for interesting new physics, even if it didn't have potential implications for the energy crisis.) Yet, some of my critics, not physicists themselves, felt that LENR didn't get anywhere near this bar. As I say, I felt that their confidence that their own view trumped that of better-qualified observers closer to the action showed a regrettable lack of epistemic humility.

I now realise that to the evidence offered in Section 3.6, I could have added the 2004 DOE report mentioned earlier. While not seen by insiders as friendly to the field, this report certainly wasn't as dismissive as the critics I have in mind. Indeed, more than 5% of the DOE assessors (i.e., one out of 18) found the evidence for LENR *completely* convincing, and a third of them found it at least partially so. These were experts called in by the DOE to examine evidence and give an opinion. Some of my critics seemed to feel that they could do better from their armchairs.

Some of these critics were linked to the Silicon Valley rationalist community, a group with an admirable commitment to epistemic self-improvement (Bay Area Rationalists, 2021). Because they lived nearby, I suggested that I could introduce them to my LENR contacts in Silicon Valley — for example, to Francis Tanzella at SRI International, or (subject to signing an NDA) to the Brillouin team. But this seemed to elicit no interest. I was reminded of a remark from Brillouin's Robert Godes, quoted in Section 3.3* (Price, 2015): 'It is sad that such people say that science should be driven by data and results, but at the same time refuse to look at the actual results' (Bjørkeng, 2015).

3.13 The Bats Come Home to Roost in Silicon Valley

My bets were settled in mid-2019. Our three judges, all physicists, agreed with my opponents that neither Brillouin nor Rossi had demonstrated evidence of LENR above 50% probability. They were more open than some of my opponents to the suggestion that the field met the lower bar, which recommended it for serious attention, given what was potentially at stake.

The Google-funded work had just been announced when the bets were finalised. I was delighted at this further evidence of interest in LENR in Silicon Valley. My opponents could take comfort from the fact that the Google team had failed to find excess heat, but from my point of view the more important thing was that they thought it worthwhile looking for it. As we saw in Section 3.9, the Google team were both well aware of, and strongly motivated by, the prudential argument for investigating LENR — the same argument that my critics had claimed to find unconvincing. Yet, this seemed to cut no ice with those of my opponents who felt that LENR failed to reach the lower bar; by their lights, the Google team had been wasting their time.

My sense that the bats are coming home to roost for cold fusion scep-tics in Silicon Valley has now received another boost. ICCF24, the 2022 meeting of the annual International Conference on Cold Fusion, is to be held in the Mountain View Museum of Computing, just up the road from the Google campus. The conference is being organised by the Anthropocene Institute (Anthropocene Institute, 2021), whose president is Carl Page. Page has been a vocal supporter of LENR, as well as of other new nuclear technologies, such as molten salt fission reactors, for several years (Page, 2016, 2019a).

In a talk from 2019 (Page, 2019a, 2019b), Page describes how he studied the claims of LENR for more than a year, initially sceptical, before agreeing to speak to Robert Godes of Brillouin Energy (in which he is now an investor). This is one of Page's slides from that talk:

LENR: Cautious view of a stigmatized field.

Scientists must test their intellectual honesty from time to time by look-ing at research with conclusions outside the consensus. Otherwise how do you know you are a scientist, and not an adherent to an ideology? Or just fashionable.

Given my interest in energy, I was asked to meet with a "cold fusion" researcher. I said "No" until I had a chance to see why the field is so unpopular with many intelligent people. After a year of reading and talk-ing to experts, I discerned a textbook example of an important and unexpected result that provoked every form of unscientific reaction, literally terrorizing honest researchers. Motivated reasoning, rampant academic nepotism, self interest, intra-disciplinary conflict, ideology, math dominance, and authoritarian rule.

I have not discussed most of the 'unscientific' factors that Page has in mind here, but it would be naive to assume that they will not be relevant to the challenges of studying the potential risks of new technologies. The field needs a handbook for combatting those sorts of factors, too.

My recommendations in Section 3.12 had a less ambitious aim: improving the culture of discussion within science, so that it does a better job of studying low-probability high-impact risks. I have offered four recommendations for the science of extreme technological risk. All of them would also be beneficial elsewhere in science, in my view.

3.14 Final Remarks on LENR

Some final remarks about the LENR case. I want to emphasise that the recent growth in interest in the field does not guarantee that cold fusion is real, let alone that it will turn out to be useful. What it does do, in my view, is to go some small way to address past failings of two kinds. The lesser failing is the lack of serious funding for the field over the past thirty years. The greater failing, responsible in large part for the lesser one, is the reputation trap.

I hope I've managed to convince some readers that in a case such as this, the reputation trap is a pathology of the scientific process. In my original *Aeon* piece, I suggested that it amounted to shooting ourselves in the foot, but this doesn't quite capture what makes it pathological. A better image would be nailing our own feet to the floor — that combines unnecessary self-harm with self-imposed impediment to the exploration of an important search space. Cold fusion will live on in the history of science as a classic example of how to get this wrong, in my view, even if it never boils a cup of tea.[6]

Acknowledgements

Many people have helped me with comments and discussion of this material at various points over the past ten years — so many, over such a long period, that I am sure that the following list is incomplete. With apologies to those who find themselves omitted, I'm grateful to everyone, named here or not, for their interest, comments, and patience: Anthony Aguirre, Shahar Avin, María Baghramian, David Bailey, Philip Ball, Jonathan Borwein, Andrew Briggs, Hasok Chang, Clive Cookson, Adrian Currie, Finnur Dellsén, Heather Douglas, Eric Drexler, Luke Drury, Larry Forsley, Russ George, Masami Hayashi, Hugh Hunt, Yasuhiro Iwamura, Brian Josephson, Jirohta Kasagi, Charlie Kennel, Adrian Kent, Ross Koningstein, Tim Lewens, Frank Ling, Michael McKubre, David Nagel, Seán Ó hÉigeartaigh, Carl Page, Ellen Quigley, Abbas Raza, Martin Rees,

---

[6]The cup of tea reference is to a line from one of the field's well-known sceptics, the physicist Robert Park (2004). McKubre replies that the field has actually boiled the equivalent of many thousands of cups of tea (Maguire, 2014). He also recounts an incident in which Park refused to look at a paper offered by someone in the field, dropping it to the floor.

Carlo Rovelli, Carl Schulman, Mikey Slezak, Alan Smith, Jaan Tallinn, Francis Tanzella, Max Tegmark, Joris van der Schot, Alyssa Vance, and Ken Wharton. I am also grateful to the editors of *Aeon* and *3QuarksDaily* for their willingness to publish my work on this topic, and to Clean Planet for their kind permission to reproduce Figure 1. Some of this work was made possible through the support of a grant on *Managing Extreme Technological Risk* from the Templeton World Charity Foundation. The opinions expressed here are my own and do not necessarily reflect the views of TWCF.

References

Angner, E. (2020). Epistemic humility — Knowing your limits in a pandemic. *Behavioural Scientist*, 13 April 2020. Available at: https://behavioral scientist.org/epistemic-humility-coronavirus-knowing-your-limits-in-a-pandemic/ (Accessed 17 December 2021).

Anonymous. (2016) Cold fusion is better left out in the cold. Editorial in *New Scientist*, 14 September 2016. Available at: https://www.newscientist.com/article/mg23130912-300-cold-fusion-is-better-left-out-in-the-cold/ (Accessed 17 December 2021).

Anonymous. (2019a). A Google programme failed to detect cold fusion — But is still a success. *Nature*, 569, 599–600. Available at: https://doi.org/10.1038/d41586-019-01675-9 (Accessed 17 December 2021).

Anonymous. (2019b). Coming in from the cold. *Nat. Mater.*, 18, 1145. Available at: https://www.nature.com/articles/s41563-019-0530-1 (Accessed 27 January 2022).

Anthropocene Institute. (2021). About the anthropocene institute. Available at: https://anthropoceneinstitute.com/about/ (Accessed 17 December 2021).

ARPA-E. (2021a). Who we are. Available at: https://arpa-e.energy.gov/about (Accessed 17 December 2021).

ARPA-E. (2021b). *Low-Energy Nuclear Reactions Workshop.* Available at: https://arpa-e.energy.gov/events/low-energy-nuclear-reactions-workshop (Accessed 17 December 2021).

Bailey, D. and Borwein, J. (2014). Fusion energy: Hope or hype? Blogpost, 23 October 2014. Available at: https://www.huffpost.com/entry/fusion-energy-hope-or-hype_b_6031968 (Accessed 17 December 2021).

Bailey, D. and Borwein, J. (2015). Cold fusion heats up: Fusion energy and LENR update. Blogpost, 28 August 2015. Available at: https://www.huffpost.com/entry/post_10010_b_8052326 (Accessed 17 December 2021).

Baldwin, M. (2015). *Making Nature: The History of a Scientific Journal.* Chicago: University of Chicago Press.

Ball, P. (2019a). Lessons from cold fusion, 30 years on. *Nature*, 569, 601. Available at: https://www.nature.com/articles/d41586-019-01673-x (Accessed 17 December 2021).

Ball, P. (2019b). Materials advances result from study of cold fusion. *MRS Bull.*, 44, 833–836. doi:10.1557/mrs.2019.260.

Bay Area Rationalists. (2021). Welcome to the Bay area rationalist community. Available at: http://www.bayrationality.com/ (Accessed 17 December 2021).

Beiting, E. (2017). Investigation of the nickel-hydrogen anomalous heat effect. The Aerospace Corporation: Report No. ATR-2017-01760. Available at: https://www.lenr-canr.org/acrobat/BeitingEinvestigat.pdf (Accessed 17 December 2021).

Berkes, H. (2012). Remembering Roger Boisjoly: He tried to stop shuttle Challenger launch. *NPR*, February 6, 2012. Available at: https://www.npr.org/sections/thetwo-way/2012/02/06/146490064/remembering-roger-boisjoly-he-tried-to-stop-shuttle-challenger-launch (Accessed 17 December 2021).

Berlinguette, C. P., Chiang, Y. M., Munday, J. N., *et al.* (2019). Revisiting the cold case of cold fusion. *Nature*, 570, 45–51. Available at: https://doi.org/10.1038/s41586-019-1256-6 (Accessed 17 December 2021).

Bjørkeng, P. K. (2015). 1 glass vann = energi til Hamar i et helt år? *Aftenposten*, 12 September 2015. Available at: https://www.aftenposten.no/norge/i/4v8q/1-glass-vann-energi-til-hamar-i-et-helt-aar (Accessed 17 December 2021).

Brooks, M. (2016). Cold fusion: Science's most controversial technology is back. *New Sci.*, 14 September 2016. Available at: https://www.newscientist.com/article/mg23130910-300-cold-fusion-sciences-most-controversial-technology-is-back/ (Accessed 17 December 2021).

Bushnell, D., Moses, R. Choi, S. (2021). Frontiers of space power and energy. NASA/TM–20210016143. Available at: https://ntrs.nasa.gov/api/citations/20210016143/downloads/NASA-TM-20210016143final.pdf (Accessed 17 December 2021).

Clean Planet. (2021). Quantum hydrogen energy — Preparing for commercial application. Available at: https://www.cleanplanet.co.jp/en/science/ (Accessed 17 December 2021).

Cookson, C. (2012). Cold fusion: A personal history. *Financial Times*, 18 August 2012. Available at: https://www.ft.com/content/4f1c41e8-e66e-11e1-ac5f-00144feab49a (Accessed 17 December 2021).

CORDIS. (2021a). Clean energy from hydrogen-metal systems. Available at: https://cordis.europa.eu/project/id/951974 (Accessed 17 December 2021).

CORDIS. (2021b). Breakthrough zero-emissions heat generation with hydrogen-metal systems. Available at: https://cordis.europa.eu/project/id/952184 (Accessed 17 December 2021).

CSER. (2017). Risk & the Culture of Science (Invite only workshop). Available at: https://www.cser.ac.uk/events/risk-the-culture-of-science-invitation-only/ (Accessed 17 December 2021).

DOE. (2004). Report of the review of low energy nuclear reactions. Available at: https://www.lenr-canr.org/acrobat/DOEreportofth.pdf. (Accessed 23 December 2021).

DOE. (2021). DOE establishes new office of clean energy demonstrations under the Bipartisan infrastructure law. Available at: https://www.energy.gov/articles/doe-establishes-new-office-clean-energy-demonstrations-under-bipartisan-infrastructure-law (Accessed 23 December 2021).

Douglas, H. (2000). Inductive risk and values in science. *Philosophy Sci.*, 67(4), 559–579. https://www.journals.uchicago.edu/doi/abs/10.1086/392855.

Douglas, H. (2007). Rejecting the ideal of value-free science. In K. Kincaid, J. Dupré, and A. Wylie (eds.), *Value-Free Science? Ideals and Illusions* (pp. 120–141). Oxford: Oxford University Press.

Douglas, H. (2009). *Science, Policy, and the Value-Free Ideal*. University of Pittsburgh Press: Pittsburgh.

Dumaine, B. (2015). This investor is chasing a new kind of fusion. *Fortune*, 28 September 2015. Available at: https://fortune.com/2015/09/27/ceo-cherokee-investment-partners-low-energy-nuclear-reaction/ (Accessed 17 December 2021).

ENEA. (2013). New advancements on the Fleischmann-Pons effect: Paving the way for a potential new clean renewable energy source? Workshop at the European Parliament, Brussels, 3 June 2013. Available through: https://www2.enea.it/en/Brussels-liason-office/news (Accessed 29 January 2024).

Fleischmann, M. and Pons, S. (1989). Electrochemically induced nuclear fusion of deuterium. *J. Electroanal. Chem. Interfacial Electrochem.*, 261, 301–308. doi.org/10.1016/0022-0728(89)80006-3.

Goodstein, D. (1994). Pariah science: Whatever happened to cold fusion? *Am. Schol.*, 63(4), 527–541. Available at: https://www.its.caltech.edu/~dg/fusion_art.html (Accessed 27 January 2022).

Greco, K. (2021). Review of 1989 and 2004 DOE reports. Available at: https://arpa-e.energy.gov/sites/default/files/2021LENR_workshop_Greco.pdf (Accessed 17 December 2021).

Habgood-Coote, J. (2018). Stop talking about fake news! *Medium*, 16 July 2018. Available at: https://medium.com/@josh_coote/stop-talking-about-fake-news-cacf90998566 (Accessed 17 December 2021).

Habgood-Coote, J. (2019). Stop talking about fake news! *Inquiry: Interdiscip. J. Philos.*, 62, 1033–1065.

Hambling, D. (2011). Success for Andrea Rossi's E-Cat cold fusion system, but mysteries remain. *Wired*, 29 October 2011. Available at: https://www.wired.co.uk/article/rossi-success (Accessed 17 December 2021).

Hasvold, J. (2018). Condensed matter nuclear science — Fiksjon eller virkelighet (fiction or reality). FFI-RAPPORT 18/00678. Available at: https://publications.ffi.no/nb/item/asset/dspace:4203/18-00678.pdf (Accessed 17 December 2021).

Hughes, N. (2020). NASA detects lattice confinement fusion. *YouTube* video, 1 August 2020. Available at: https://www.youtube.com/watch?v=e2pcrgFb7L4 (Accessed 17 December 2021).

ICCF-24. (2021). *ICCF-24 Silicon Valley.* Available at: https://www.iccf24.org/ (Accessed 17 December 2021).

Jones, S., Palmer, E., Czirr, J., *et al.* (1989). Observation of cold nuclear fusion in condensed matter. *Nature*, 338, 737–740. https://doi.org/10.1038/338737a0.

Kaneko, K. (2016). Successful reproduction of patents in the US accelerates re-evaluation of "cold fusion". *Nikkei*, 9 September 2016. Available at: https://www.nikkei.com/article/DGXMZO06252800Z10C16A8000000/ (Accessed 23 December 2021 via Google Translate).

Kasagi, J. and Iwamura, Y. (2008). Country history of Japanese work on cold fusion, in *ICCF-14 International Conference on Condensed Matter Nuclear Science*, Washington, DC. Available at: https://www.lenr-canr.org/acrobat/KasagiJcountryhis.pdf (Accessed 11 January 2022).

Kitamura, A., Takahashi, A., Takahashi, K., Seto, R., Hatano, T., *et al.* (2018). Excess heat evolution from nanocomposite samples under exposure to hydrogen isotope gases. *Int. J. Hydrogen Energy*, 43(33), 16187–16200. Available at: https://doi.org/10.1016/j.ijhydene.2018.06.187.

Koningstein, R. and Fork, D. (2014). What it would really take to reverse climate change. *IEEE Spectrum*, 18 November 2014. Available at: https://spectrum.ieee.org/what-it-would-really-take-to-reverse-climate-change (Accessed 17 December 2021).

Koziol, M. (2020). Spacecraft of the future could be powered by lattice confinement fusion — NASA researchers demonstrate the ability to fuse atoms inside room-temperature metals. *IEEE Spectrum*, 5 August 2020. Available at: https://spectrum.ieee.org/nuclear-fusiontokamak-not-included (Accessed 17 December 2021).

Koziol, M. (2021). Whether cold fusion or low-energy nuclear reactions, U.S. Navy researchers reopen case. *IEEE Spectrum*, 22 March 2021. Available at: https://spectrum.ieee.org/cold-fusion-or-low-energy-nuclear-reactions-us-navy-researchers-reopen-case (Accessed 17 December 2021).

Kuhn, T. (1962). *The Structure of Scientific Revolutions.* Chicago: University of Chicago Press.

LENR-CANR.ORG. (2021). LENR-CANR.ORG — A library of papers about cold fusion. Available at: https://lenr-canr.org/ (Accessed 17 December 2021).

Levi, G., Foschi, E., Hartman, T., Höistad, B., Petterson, R., Tegnér, L., and Essén, H. (2013). Indication of anomalous heat energy production in a reactor device. Available at: https://arxiv.org/abs/1305.3913 (Accessed 17 December 2021).

Levi, G., Foschi, E., Höistad, B., Petterson, R., Tegnér, L., and Essén, H. (2014). Observation of abundant heat production from a reactor device and of isotopic changes in the fuel. Preprint. Available at: http://www.sifferkoll.se/sifferkoll/wp-content/uploads/2014/10/LuganoReportSubmit.pdf (Accessed 17 December 2021).

Lewan, M. (2015a). Replication attempts are heating up cold fusion. Blogpost, 1 February 2015. Available at: https://animpossibleinvention.com/2015/02/01/replication-attempts-are-heating-up-cold-fusion/ (Accessed 17 December 2021).

Lewan, M. (2015b). Rossi has been granted US patent on the E-Cat — Fuel mix specified. Blogpost, 25 August 2015. Available at: https://animpossible invention.com/2015/02/01/replication-attempts-are-heating-up-cold-fusion/ (Accessed 17 December 2021).

Lewan, M. (2015c). Swedish scientists claim LENR explanation break-through. Blogpost, 15 October 2015. Available at: https://animpossibleinvention. com/2015/10/15/swedish-scientists-claim-lenr-explanation-break-through/ (Accessed 17 December 2021).

Lucretius. (1916). *De rerum natura*. Translated by William Ellery Leonard. New York: E. P. Dutton.

Lundin, R. and Lidgren, H. (2015). Nuclear spallation and neutron capture induced by ponderomotive wave forcing. *IRF Sci. Res.*, 305. ISSN 0284-1703. Available at: https://www.semanticscholar.org/paper/Nuclear-Spallation-and-Neutron-Capture-Induced-by-Lundin-Lidgren/19d09a644d8190ba899b 0e7588341a4787c27246 (Accessed 17 December 2021).

Maguire, J. (2014). Dr. Michael McKubre: Experimental cold fusion, pseudo-skepticism, and progressing LENR. *Q-Niverse Podcast*, 25 April 2014. Available at: https://www.podomatic.com/podcasts/jmag0904/episodes/2014-04-24T11_40_44-07_00 (Accessed 17 December 2021).

Malle, L. (1981). *My Dinner with Andre* (Movie). Saga Productions.

Marwan, J., McKubre, M., Tanzella, F., Hagelstein, P., Miles, M, Schwarz, M., *et al.* (2010). A new look at low-energy nuclear reaction (LENR) research: A response to Shanahan. *J. Environ. Monit.*, 12, 1765–1770. Available at: http://dx.doi.org/10.1039/c0em00267d.

McDonough, W. (2015). Investing in innovation for the common good. Webpage. Available at: https://mcdonough.com/writings/investing-innovation-common-good/ (Accessed 17 December 2021).

McKay, P. (2017). Disagreement in science and beyond — A workshop organised by WEXD (Dublin) and CSER (Cambridge). Available at: http://when expertsdisagree.ucd.ie/disagreement_in_science_and_beyond/ (Accessed 17 December 2021).

McKubre, M. (2009). Cold fusion (LENR) — One perspective on the state of the science. *15th International Conference on Condensed Matter Nuclear Science*. Rome, Italy: ENEA. Available at: https://www.lenr-canr.org/acrobat/McKubreMCHcoldfusionb.pdf (Accessed 17 December 2021).

McKubre, M. (2019). Critique of *Nature* perspective article on google-sponsored Pd-D and Ni-H experiments. *Infinite Energy*, 146, July/August 2019, 1–4. Available at: https://www.infinite-energy.com/iemagazine/issue146/McKubreGoogle.pdf (Accessed 17 December 2021).

Metzler, T. (2021). Towards a LENR reference experiment. Workshop presentation, 21 October 2021. Available at: https://www.youtube.com/watch?v=Ec9OnfWvOjs&t=1726s (Accessed 29 December 2021).

Miura. (2021). MIURA CO. LTD. and Clean Planet Inc. conclude an agreement for joint development of industrial boilers that use quantum hydrogen energy, 28 September 2021. Available at: https://www.miuraz.co.jp/news/newsrelease/2021/1132.php (Accessed 17 December 2021).

Mizuno, T. (2017). Observation of excess heat by activated metal and deuterium gas. *J. Condensed Matter Nucl. Sci.*, 25, 1–25. https://www.lenr-canr.org/acrobat/MizunoTpreprintob.pdf.

NASA Glenn Research Center. (2021). Lattice confinement fusion. 15 April 2020. Available at: https://www1.grc.nasa.gov/space/science/lattice-confinement-fusion/ (Accessed 17 December 2021).

NikkeiBP. (2021). Boilers using "nuclear fusion/heat" are put into practical use, and heat is taken out with metal laminated chips. Jointly developed by Miura Co., Ltd. and Clean Planet, commercialized in 2023. 4 October 2021. Available at: https://project.nikkeibp.co.jp/ms/atcl/19/news/00001/02043/?ST=msb&P=2 (Accessed 17 December 2021 via Google Translate).

NYU Philosophy. (2021). NYU guidelines for respectful philosophical discussion. Available at: https://as.nyu.edu/content/nyu-as/as/departments/philosophy/climate/initiatives/nyu-guidelines-for-respectful-philosophical-discussion.html (Accessed 17 December 2021).

Page, C. (2016). Low energy nuclear reactions work and could supplant fossil fuels. *Edge* 2016. Available at: https://www.edge.org/response-detail/26753?fbclid=IwAR3iKDxvjbnZlo1FEcyrnvbQK6F4AHUiXfCfhTO_CucuxosR-kn03n9CKOY (Accessed 17 December 2021).

Page, C. (2019a). Context and thoughts on LANR/LENR. Slide presentation, 24 March 2019. Available at: https://www.lenr-canr.org/acrobat/PageCcontexttho.pdf (Accessed 17 December 2021).

Page, C. (2019b). Context and thoughts on LANR/LENR. *YouTube*, 16 May 2019. Available at: https://www.youtube.com/watch?v=Oebkw0G7tEw (Accessed 17 December 2021).

Park, R. (2004). Cold fusion: Just when you think life can't get any sillier. *What's New*, 2 April 2004. Available at: http://bobpark.physics.umd.edu/WN04/wn080604.html (Accessed 17 December 2021).

Pines, V., Pines, M., Chait, A., Steinetz, B. M., Forsley, L. P., Hendricks, R. C., *et al.* (2020). Nuclear fusion reactions in deuterated metals. *Phys. Rev. C*, 101(4), 044609. DOI: 10.1103/PhysRevC.101.044609.

Price, H. (2015). Why do scientists dismiss the possibility of cold fusion? *Aeon*, 21 December 2015. Available at: https://aeon.co/essays/why-do-scientists-dismiss-the-possibility-of-cold-fusion (Accessed 17 December 2021).

Price, H. (2016). Is the cold fusion egg about to hatch? *Aeon*, 24 March 2016. Available at: https://aeon.co/ideas/is-the-cold-fusion-egg-about-to-hatch (Accessed 17 December 2021).

Price, H. (2019). Icebergs in the room? Cold fusion at 30. *3 Quarks Daily*, 4 March 2019. Available at: https://3quarksdaily.com/3quarksdaily/2019/03/icebergs-in-the-room-cold-fusion-at-thirty.html (Accessed 17 December 2021).

Reich, E. (2011). Speedy neutrinos challenge physicists. *Nat. News*, 477(7366), 520. doi:10.1038/477520a.

Reich, E. (2012). Flaws found in faster-than-light neutrino measurement. *Nat. News*, doi:10.1038/nature.2012.10099.

Rhodes, R. (1986). *The Making of the Atomic Bomb*. New York: Simon and Schuster.

Russell, E. (2018). Francis Tanzella on the cold fusion now! Podcast. *Cold Fusion Now!* 13 October 2018. Available at: https://coldfusionnow.org/francis-tanzella-on-the-cold-fusion-now-podcast/ (Accessed 17 December 2021).

Salamon, M., Wrenn, M., Bergeson, H., *et al.* (1990). Limits on the emission of neutrons, γ-rays, electrons and protons from Pons/Fleischmann electrolytic cells. *Nature*, 344, 401–405. doi.org/10.1038/344401a0.

Scott, R. (2015). *The Martian* (Movie). Scott Free Productions.

Steinetz, B., Benyo, T., Chait, A., Hendricks, R., Forsley, L., Baramsai, B., *et al.* (2020). Novel nuclear reactions observed in bremsstrahlung-irradiated deuterated metals. *Phys. Rev. C*, 101(4), 044610. DOI: 10.1103/PhysRevC.101.044610.

Szpak, S., *et al.* (2008). SPAWAR systems Center-Pacific Pd:D Co-deposition research: Overview of refereed LENR publications. In *ICCF-14 International Conference on Condensed Matter Nuclear Science*, Washington, DC. Available at: https://www.researchgate.net/publication/242327687_SPAWAR_Systems_Center-Pacific_PdD_CoDeposition_Research_Overview_of_Refereed_LENR_Publications. (Accessed 17 December 2021).

Takahashi, A., Kitamura, A., Takahashi, K., Seto, R., Matsuda, Y., Iwamura, Y., *et al.* (2017). Phenomenology and controllability of new exothermic reaction between metal and hydrogen. *Brief Summary Report of MHE Project Japan for 2015 October–2017 October*. Available at: https://www.researchgate.net/publication/322160963_Brief_Summary_Report_of_MHE_Project_Phenomenology_and_Controllability_of_New_Exothermic_Reaction_between_Metal_and_Hydrogen (Accessed 17 December 2021).

Tanzella, F. (2016). *Isoperibolic Hydrogen Hot Tube Reactor Studies: Interim Progress Report for the Period 1 March–5 December 2016*. SRI International

Project P21429. https://brillouinenergy.com/newwebsite/wp-content/uploads/2018/12/SRI_ProgressReport.pdf (Accessed 17 December 2021).

Tanzella, F. (2018a). *Isoperibolic Hydrogen Hot Tube Reactor Studies: Technical Progress Report for the Period 1 January–31 December 2017.* SRI International Project P21429. Available at: https://brillouinenergy.com/newwebsite/wp-content/uploads/2018/12/SRI_Technical_Report.pdf (Accessed 17 December 2021).

Tanzella, F. (2018b). *Isoperibolic Hydrogen Hot Tube Reactor Studies: Final Progress Report for the July 1st 2016 Through December 31st 2018.* SRI International Project P21429. Available at: https://brillouinenergy.com/newwebsite/wp-content/uploads/2019/04/Brillouin-SRI-Technical-Progress-Report-Final-Public-2018.pdf (Accessed 17 December 2021).

UBC. (2021). Curtis Berlinguette. Available at: https://groups.chem.ubc.ca/cberling/curtis-berlinguette/ (Accessed 17 December 2021).

Vineberg, S. (2016). Dutch book arguments. In *The Stanford Encyclopedia of Philosophy*, E. Zalta (ed.). Available at: https://plato.stanford.edu/archives/spr2016/entries/dutch-book/ (Accessed 17 December 2021).

Wang, B. (2015). China's LENR is getting excess 600 watts of heat from 780 watts of input power. Blogpost, 8 June 2015. Available at: https://www.nextbigfuture.com/2015/06/chinas-lenr-is-getting-excess-600-watts.html (Accessed 17 December 2021).

Wells, D. P., Campbell, R., Chase, A., Daniel, J., Darling, M., Green, C., *et al.* (2014). Low energy nuclear reaction aircraft — 2013 ARMD seedling fund phase I project. NASA/TM–2014-218283. Available at: https://nari.arc.nasa.gov/sites/default/files/Wells_TM2014-218283%20Low%20Energy%20Nuclear%20Reaction%20Aircraft_0.pdf (Accessed 17 December 2021).

Wikipedia Contributors. (2021a). Sergio Focardi. In *Wikipedia, the Free Encyclopedia.* Available at: https://en.wikipedia.org/w/index.php?title=Sergio_Focardi&oldid=997654571 (Accessed 22 December 2021).

Wikipedia Contributors. (2021b). Cold fusion. In *Wikipedia, the Free Encyclopedia.* Available at: https://en.wikipedia.org/w/index.php?title=Cold_fusion&oldid=1059561939 (Accessed 10 December 2021).

Wikipedia Contributors. (2021c). Space shuttle challenger disaster. In *Wikipedia, the Free Encyclopedia.* Available at: https://en.wikipedia.org/w/index.php?title=Space_Shuttle_Challenger_disaster&oldid=1059690149 (Accessed 11 December 2021).

Wikipedia Contributors. (2021d). Ad hominem. In *Wikipedia, the Free Encyclopedia.* Available at: https://en.wikipedia.org/w/index.php?title=Ad_hominem&oldid=1061366737 (Accessed 30 December 2021).

Zeldovich, L. (2019). A history of human waste as fertilizer. *JSTOR Daily*, 18 November 2019. Available at: https://daily.jstor.org/a-history-of-human-waste-as-fertilizer/ (Accessed 17 December 2021).

https://doi.org/10.1142/9781800614826_0004

Chapter 4

Foreseeing Extreme Technological Risk

Luke Kemp

4.1 Introduction

In 1945, the first atomic weapon was detonated in a deserted field in Nevada. The Trinity Test marks the beginning of perhaps the most extreme technological risk. It is also a lesson. It was a largely unforeseen and abrupt advance. The regulation of nuclear weapons lagged for decades thereafter. After the failed Baruch Plan of 1946, it took another two decades for effective international management to be assembled with the 1963 Limited Nuclear Test Ban Treaty and the later Strategic Arms Limitation Talks (SALT I and II) processes. By the time of SALT II, nuclear weapons stockpiles had swelled to over 53,000 warheads.

What would have happened if the threat of nuclear weapons had been foreseen decades in advance and taken seriously by policymakers? Even for less extreme risks, such as lives lost in automobile accidents prior to the introduction of seatbelts, anticipation and foresight can save lives. This is why anticipatory governance is often flagged as a hallmark of responsible innovation and science (Stilgoe *et al.*, 2013; Owen *et al.*, 2012; Flachsland *et al.*, 2009).

Extreme risks are potentially adverse events that are high impact and low probability. Such risks are generally more difficult to foresee and have severe and frequently long-lasting impacts. They are often referred to as 'black swans': rare and impactful events that are predicted retrospectively but rarely prospectively (Taleb, 2007a). Preparing for such outliers is a devilish problem that requires innovative ways of thinking about the future (Goodwin and Wright, 2010). It should minimally require making resilient systems, or, more ambitiously, constructing anti-fragile systems that can gain from shocks (Taleb, 2012). The project on Managing Extreme Technological Risk (METR) at the Centre for the Study of Existential Risk (CSER) sought to develop such foresight methods for severe risks arising from emerging technologies, particularly advanced biotechnologies.

In this chapter, we present some of the key lessons learned from this project, summarise useful approaches that we refined in horizon scanning and evidence synthesis, and advance some ideas on designing foresight systems for anticipating extreme risks.

4.1.1 *Extreme betting and extreme foresight*

Forecasting and foresight are frequently confused, and this includes their use by scholars of extreme risks. The two are functionally distinct, with differing strengths and weaknesses. They also have different applicability to the study of low-probability, high-impact events, particularly over the long term. Common techniques for both foresight and forecasting are summarised in Table 1.

Table 1. A summary of common foresight and forecasting techniques.

Class	Foresight/Forecasting Technique
Judgemental	Delphi technique
	Gaming and role playing (foresight specific)
	Literature reviews
	Prediction markets (forecasting specific)
	Scenario construction (foresight specific)
	Superforecasters (forecasting specific)
	Surveys
Statistical	Statistical models (forecasting specific)

4.1.2 *Forecasting*

Forecasting focuses on making probabilistic (usually numerical) estimates of bounded events. This is usually confined to what are called 'binary bets': predictions between two or slightly more well-defined outcomes (Taleb and Tetlock, 2013), for example, picking an Oscar winner, the next President, or the likelihood of war casualties exceeding a given threshold in a particular area.

Forecasting is ultimately about probability; it says very little about the impact of the bet being true or not. This places a high burden on asking the right questions and requires further work to understand how these fit into a wider picture of the future.

For forecasting, the most effective and reliable methods have proven to be the use of superforecasters, prediction markets, and different versions of the Delphi method. Superforecasting is now perhaps the most well-known approach thanks to the success of the 2015 book *Superforecasters: The Art and Science of Prediction* by Phillip E. Tetlock and Dan Gardner. In it, they detail what makes a good forecaster and how to combine individual forecasters into groups that are even more accurate (Tetlock and Gardner, 2016). For forecasting, the whole can be greater than the sum of the parts. In short, it involves measuring the accuracy of forecasters with a Brier score (a common measurement for the accuracy of past predictions), aggregating them into small teams that share information, and then extremising[1] their predictions with a machine learning algorithm. Tetlock's work, alongside numerous others, spurred a wealth of literature on the most successful attitudes (Mellers *et al.*, 2019), psychological strategies (Mellers *et al.*, 2014), and other attributes of effective forecasters.

Tetlock, along with decision scientists Barbara Mellers and Don Moore, famously used the superforecasters approach to win a prediction tournament hosted by the Intelligence Advanced Research Projects Agency (IARPA), the primary research body for US intelligence agencies.

---

[1]Slightly increasing probabilities that are above 50% and decreasing those below 50%. In other words, when the weighted average of the best forecasters was biased in one direction, it was pushed further. For instance, a 30% probability could be extremised to 15% and an 85% probability to 70%.

They beat numerous research and industry teams, including those using the Delphi method, prediction markets, and other tools.

Since then, forecasting tools, and superforecasting in particular, have become favoured in the communities surrounding the study of existential and extreme risks, such as the Effective Altruism (EA) movement. Online platforms such as Metaculus (2021) regularly cover speculative questions concerning existential risks, such as the likelihood that a synthetic biological weapon will infect 100 people by 2030, whether a future global catastrophe will be due to biotechnology or bioengineered organisms, and the time it will take for artificial general intelligence (AGI) to transform into superintelligence. Similarly, well-cited studies on foreseeing the development of AGI have relied on the gathering of forecasts by relevant experts through surveys (Grace *et al.*, 2018).

It is unlikely that the success of forecasting, super or otherwise, can be easily achieved for extreme risks. Forecasting, while a helpful and well-honed tool, is not currently well suited to hunting black swans. This is due to several points of mismatch between the strengths of superforecasting and the characteristics of catastrophic risks.

First, as Tetlock notes, superforecasters have only been effective for predicting short-term events within a year. By contrast, most extreme technological and environmental risks operate over long timescales.

Second, forecasting works best for restricted binary bets. As noted earlier, this relies on selecting the right questions, a formidable task for which forecasting has no agreed answer, and one which is critical for extreme risks.

Third, for any given risk, there is an enormous spread of potential outcomes. Even if forecasting can appropriately capture the likelihood of tail risk, we would still need to run forecasting tools against an enormous number of binary bets across the distribution. Accurately assessing the wrong bet in the spread of outcomes can be disastrous. Morgan Stanley successfully predicted the 2007 sub-prime mortgage crisis and took a binary hedge to prepare. It had one accurate forecast but failed to predict how deep the resulting financial crisis would be. It lost billions (Taleb and Tetlock, 2013). Forecasting methods have been applied to continuous payoff issues. This includes forecasting wide baskets of plausible fatalities from COVID-19 to create a probability distribution. Such an approach is less frequently used and still runs against the other problems of forecasting outlined in this section.

Fourth, forecasting has never been applied to the domain of tail risks. No forecasts have ever been run to determine the effectiveness of judgements between 0.000001 and 0.1%. It is unclear how exactly one would run such a tournament, or how frequently it would need to be done. Like running forecasting bets across the range of plausible outcomes, such an approach seems challenging, if not unfeasible.

Fifth, forecasting focuses on well-calibrated predictions, while decisions about the future ultimately hinge as much on consequences and wise weighing of benefits, risks, and probabilities. Imagine two forecasters in a casino. One is incredibly well calibrated and confident, with predictions that are right 99% of the time. In the 1% of the time they are wrong, they have unfortunately gone all in. They go home bankrupt. The second forecaster is a professional pessimist. They are less well calibrated and take a precautionary approach, losing many bets by erring on the side of caution, but ultimately going home with a modest return. The second investor, while less well calibrated, is more vigilant to tail risk. They take a precautious approach that is sensitive to outcomes at the expense of probabilistic accuracy. A high Brier score is no guarantee for survival.[2] That said, this is matter of strategy, not a weakness of superforecasting per se. Well-calibrated forecasters combined with an appropriately risk-averse strategy will trump poor forecasters with the same approach. This is a warning that even if high forecasting accuracy is achieved, it is no guarantee of effective action.

In short, being a good forecaster in a binary space does not mean good actual performance in the real world of continuous payoffs, especially with nonlinearities (Taleb, 2020). Forecasting is currently neither designed nor well suited for extreme risks. It is best used for domains such as sports and elections, not wars and revolutions. In more technical language, it is most appropriate for binary, not continuous payoffs (Taleb, 2020). Indeed, science is about understanding properties, not single-point estimates. Risk management is about guarding against extremes, not the average (Taleb *et al.*, 2020). Predictions alone provide little insight if they are not coupled with transparent reasoning and deliberation. This is a point already made by others for forecasting AGI (Cremer, 2021). Expert predictions for such sophisticated technologies often display deep disagreements. They can say little about the pathways towards a given outcome, the right questions

---

[2] This thought experiment is modified from a similar story in Taleb's *Fooled By Randomness* (Taleb, 2007b).

to ask, or what early warning signs to be vigilant for (Cremer and Whittlestone, 2020).

One recent pre-print has attempted to tackle some of the limitations of applying forecasting to existential risk. 'Improving Judgments of Existential Risk', authored by a team involving Tetlock, suggests using reciprocal scoring (effectiveness at guessing the predictions of others) as a proxy for longer-term accuracy and applying this to existential-risk-relevant questions (Karger *et al.*, 2022). While this is a welcome and helpful step forward, it does not address most of the criticisms covered in this chapter. Moreover, reciprocal scoring is an unproven proxy for an area like extreme risks. Nonetheless, the attempts to combat the decay of forecaster accuracy over longer timespans and to pinpoint early warning signals which forecasters can monitor should be applauded. These are vital steps forward.

Forecasting should not be abandoned when dealing with catastrophic risks. Instead, we need a sober acknowledgement of its strengths and weaknesses. It has a role in understanding and planning for catastrophic risks. Yet, this is likely not as central a role as it has played to date. Forecasting will likely be more effective as a tool used for assessing early warning signals that are identified and updated by a larger foresight exercise. Rather than viewing probabilistic forecasting and judgemental foresight as a dichotomy, we should move towards integrating the two. They are, after all, complementary (Scoblic and Tetlock, 2020; Tetlock and Scoblic, 2021). The same holds for forecasting and extreme value theory. Vigilance towards systemic risk can help achieve proper levels of attention towards potential catastrophes. On the other hand, forecasting can help guide policy choices, examine early warning signals, and ensure that society is not overly risk averse (Tetlock *et al.*, 2022). This is the most pressing challenge for the foresight of extreme risks: finding the right mix of forecasting, foresight, and precaution.

4.1.3 *Foresight for understanding extreme risk*

While definitions of foresight vary, it differs substantially from forecasting. Most definitions refer to it roughly as a set of methods to create plausible visions of the future and integrate this understanding into current decision-making (Cuhls, 2003; Martin, 2010). While both are concerned about the future, foresight ultimately seeks to paint the parameters of plausible futures, rather than place discrete predictions about specific events. This does not mean that the two are mutually exclusive. Foresight encapsulates an array of techniques including different versions of the Delphi technique, scenarios, literature reviews, gaming and role-playing exercises, surveys, and combinations of these.

While foresight is a well-developed field, it has an Achilles heel. We do not have a refined study of the track record of different approaches. In short, we do not have enough studies on what works. For instance, there are summaries of different approaches to scenario building (Amer *et al.*, 2013), but seemingly no equivalent study on what has been shown to work. Much of this is due to understandable conceptual problems. Foresight studies frequently look to either loosely identify issues on the horizon of current thinking or trace out future scenarios. Scenarios and issues cannot be boiled down to a simple yes/no response. Another problem is that foresight studies frequently cast decades into the future. There is often a lack of incentives to review whether such approaches were correct, and many recent methods simply have not existed long enough to review their long-term efficacy.

Many foresight studies, such as scenarios, also provide multiple differing depictions of the future. If only one of these (or elements of one) comes to fruition, then the probabilistic track record of the exercise will inevitably seem poor. This would of course be a misunderstanding: if the exercise effectively identified a realistic scenario and helped decision-makers prepare for it, then it can be largely considered a success, even if it was ultimately only one scenario out of many that did not occur.

Finally, the act of conducting foresight can change the future. This may even be the aim: to identify a future risk and prevent it from occurring, or to locate a neglected opportunity and capitalise upon it. There can be a tragedy to this. Imagine a world in which a US senator is involved in a scenario role-playing exercise in January of 2000. In it, terrorists brutally hijack a plane and turn it into a high-speed missile aimed at the White House. The shocked Senator spearheads legislation for reinforced cockpits. It is passed by early 2001. 19 conspirators hear of the new regulations and abandon plans to hijack multiple flights later that year. 9/11 never occurs, but ultimately no one even knows it could have occurred. The Senator is ridiculed for the precautious and costly measures. Everyone loves a firefighter, but no one loves a safety inspector.

Heroes usually stand on deeds of reaction, not pre-emption. The problem is already well known in disaster risk management. Leaders face sinister incentives. It is far more politically popular to respond to a disaster than to pass the often-expensive measures to prevent one (Kunreuther *et al.*, 2016; Neumayer *et al.*, 2014). This poses a particularly treacherous problem for extreme risks. The impacts can be so severe that responding after calamity strikes would be a fool's errand.

For some, the purpose of foresight is not necessarily to produce accurate depictions of the future. Scenarios, games, and role playing are often more focused on having participants cultivate a forward-looking and open mindset in an interactive way. They may be used to simply have decision-makers consider potential issues that are otherwise unforeseen. For example, Operation Dark Winter in 2001 was a high-level scenario of a bioterrorist attack in the US. High-ranking US officials acted as government officials in a simulated smallpox attack on the US. The activity aimed to both promote government and public awareness and identify weaknesses and improvements in prevention and response strategies. Unfortunately, there are no guarantees that lessons will be learned or integrated into policy. Nonetheless, such simulations can be valuable even if the most likely or impactful issues are not brought to the fore. Yet, in most cases, an accurate understanding of the future is beneficial, if not necessary. Decision-making entails opportunity costs as preparing for all potential futures is not possible and focusing on ones that are implausible is a waste. Whether it be stress testing or crafting interventions, having an accurate comprehension of plausible and impactful futures is needed.

These limitations do not mean that we are collectively blind when it comes to the effectiveness of foresight tools. Some studies do exist. For instance, one review by an expert panel of a long-term Delphi study on developments in the health sector found that 14/18 issues identified by participants could be considered accurate (Parente and Anderson-Parente, 2011). Similarly, a decade-long review of a 2009 global conservation horizon scan found that 6 of the 15 identified issues had moderate effects and five had become widely known and impactful, such as cultured meat (Sutherland *et al.*, 2019). One recent refinement of the Delphi method, known as the Investigate, Discuss, Estimate, Aggregate (IDEA) Protocol, has outperformed other forms of structured expert elicitation and prediction marks in forecasting tournaments (Hanea *et al.*, 2017, 2018), as well as demonstrating some short-term success in anticipating issues in bioengineering (Wintle *et al.*, 2017; Kemp *et al.*, 2020).

Foresight is frequently opened with horizon-scanning processes, some of which have a promising track record. Horizon scanning refers to an array of foresight techniques which look to identify emerging issues on the margins of current thinking (Könnölä *et al.*, 2012; Schultz, 2006). This can include 'exploratory' horizon scanning for new issues (such as a novel technology) and 'issues-centred' horizon scanning which track signals which can herald the emergence of an issue (Amanatidou *et al.*, 2012). These tend to occur early in the foresight process before any actions are decided. Horizon scanning through structured expert elicitation now has a long and illustrious track record in global conservation biology (Sutherland *et al.*, 2019; Sutherland and Woodroof, 2009), as well as success in influencing Antarctic and Southern Ocean research priorities (Kennicutt *et al.*, 2015; Kennicutt *et al.*, 2019).

Our approach to managing extreme technological risk deliberately focused on foresight. While it may not be a perfect tool, it is a better fit for purpose than forecasting methods. Most foresight methods do not have a sufficiently reliable record of their effectiveness, yet there are enough enticing avenues that we can begin to apply and refine tools for extreme technological risk. In METR, we made use of some of the promising tools mentioned earlier, such as horizon scanning and the IDEA protocol.

4.2 Horizon Scans

Our horizon scans acted as a kind of 'wisdom of the crowds' among experts. We aggregated the opinions of diverse expert groups who truthfully shared information and deliberated to come to better judgements. They were exercises in expert deliberative democracy. The IDEA protocol is a particularly successful form of the Delphi method of expert elicitation that has been applied to topics ranging from pollinator abundance (Barons *et al.*, 2018) and natural resource management (Hemming *et al.*, 2018a) to biosecurity (Kemp *et al.*, 2021). In it, experts put forward issues that they see as emerging, impactful, and plausible. These are then collectively and anonymously scored. The experts then deliberate over the issues and thereafter rescore them. The rescored issues are ranked once again, resulting in a final list of 10–20 top issues. These are priority topics that can be the focus of public deliberation and action (Hemming *et al.*, 2018b).

We conducted two horizon scans in bioengineering in 2017 (Wintle *et al.*, 2017) and 2020 (Kemp *et al.*, 2020). Both focused on risks and

Table 2.   The top 20 identified issues in the 2020 bioengineering horizon scan.

<5 Years	5–10 Years	>10 Years
Access to biotechnology through outsourcing	Agricultural gene drives	Bio-based production of materials
Crops for changing climates	Neuronal probes expanding new sensory capabilities	Live plant dispensers of chemical signals
Function-based design in protein engineering	Distributed pharmaceutical development and manufacturing	**Malicious use of advanced neurochemistry**
Philanthropy shapes bioscience research agendas	Genetically engineered phage therapy	Enhancing carbon sequestration
State and international regulation of DNA database use	Human genomics converging with computing technologies	Porcine bioengineered replacement organs
	Microbiome engineering in agriculture	The governance of cognitive enhancement
	Phytoremediation of contaminated soils	
	Production of edible vaccines in plants	
	The rise of personalised medicine such as cell therapies	

opportunities emerging from the field over the coming decades. Our aim in the horizon scan was to identify impactful, novel, and plausible issues on the margins of current thinking, and to use aggregated expert judgement to spot emerging technologies and separate the hype from reality. Tables 2 and 3 summarise the top identified issues of both horizon scans. The technologies that could plausibly directly constitute extreme risks have been highlighted in bold font. These scans did not just focus on extreme risks. A broader view was taken to both ensure a wider set of experts and to help gain information not just about risks but also opportunities and possible interventions. After all, risk assessment needs more than just knowledge of threats.

We used a similar approach in collaboration with the World Health Organization (WHO) for dual-use research of concern (DURC), that is,

Table 3.   The top 20 identified issues in the 2017 bioengineering horizon scan.

<5 Years	5–10 Years	>10 Years
Artificial photosynthesis and carbon capture for producing biofuels	Regenerative medicine: 3D printing body parts and tissue engineering	New makers disrupt pharmaceutical makers
Enhanced photosynthesis for agricultural productivity	Microbiome-based therapies	Platform technologies to address emerging disease pandemics
New approaches to synthetic gene drives	Producing vaccines and human therapeutics in plants	**Challenges to taxonomy-based description and management of biological risk**
Human genome editing	Manufacturing illegals drugs using engineered organisms	Shifting ownership models in biotechnology
Accelerating defence agency research in biological engineering	Reassigning codons as genetic firewalls	Securing the critical infrastructure needed to deliver the bioeconomy
	Rise of automated tools for biological design, test, and optimisation	
	Biology as information science: Impacts on global governance	
	Intersection of information security and bio-automation	
	Effects of the Nagoya Protocol on biological engineering	
	Corporate espionage and biocrime	

advances in the life sciences that could be easily misapplied to cause harm. The 15 priority issues identified in the scan are summarised in Table 4.

We took an evolutionary approach to the horizon scans, making small, targeted modifications to the approach of the second scan. This involved changes to the participant pool, the definition of bioengineering, and the

Table 4.   The priority 15 issues identified by the WHO DURC horizon scan.

Timeframe	Issue
5 years	Bioregulators
	Cloud laboratories (outsourcing biotechnology)
	De Novo synthesis of variola
	Research on SARS-CoV-2: Pathogenesis, host range, and cell tropism
	Synthetic genomics platforms for virus reconstruction, evolution, and engineering
5–10 years	Identification of novel biological constructs with deep learning
	Extreme high-throughput discovery systems
	Gain of function experiments in vectors
	Stabilised biological and toxin particles for compound delivery
	Targeted gene drive applications
10+ years	Hostile exploitation of neurobiology
	Nanotechnology and nanoparticle toxicity
Governance issues	Infodemic, misinformation, disinformation, and DURC
	The impact of a continued lack of a global DURC framework
	Implementing safety by design in dual-use research projects

ability to address black swan events. We used a larger pool of participants, increasing the number of participants from 27 to 38. We also drew on a far more diverse participant pool with participants from 6 continents and 13 countries, compared to a Transatlantic scan in 2017 which focused on US and UK participants. This effectively developed the horizon scan from a transatlantic project into a more global endeavour.

We also moved towards a wider definition. Our definition in both the 2017 and 2020 scans referred to applying techniques and ideas from engineering to biological systems. The critical difference was that our 2017 scan excluded implants, while our 2020 exercise did not, leading to the latter capturing issues such as expanding the scope to include a wider set of technologies including implants and brain–machine interfaces (Neuronal Probes Expanding New Sensory Capabilities).

Another change was the deliberate use of a method to better identify low-probability, high-impact black swan events. This was the inclusion of the devil's advocates. Scholars have noted that most existing foresight and forecasting techniques are inadequate for finding rare, high-impact events, but the inclusion of 'exceptional individuals' who challenge the emerging group consensus could do so (Goodwin and Wright, 2010). In the first step of the expert elicitation process, we empowered two individuals to put forward more speculative and impactful topics. In the deliberation phase, we asked two separate individuals to intentionally act as defenders of more outlandish ideas and to challenge the group consensus on different issues. This appears to have been successful. Out of nine issues proposed by the devil's advocates in the initial round, six made it through to the second round of four. Four made it into the final priority list of 20. Given the number of participants (38), we would only expect every second participant to have contributed an issue that made it to the final 20. The devil's advocates averaged two each.

The WHO DURC scan was also subject to experimentation. This included providing a common template into which participants could draft their issues (for efficiency and comparability), giving participants basic training and tips to improve their judgement, and providing a 'signals package' of relevant research synthesis platforms and previous foresight exercises. These modifications were intended to gift the experts a better shared base of information, improve their judgement, and provide more comparable and comprehensive issues.

Our evolutionary approach to foresight is about honest reflection. We do not know what the most fit-for-purpose foresight tools are when it comes to extreme risks. Our process of informed tinkering, and trial and error, looks to constantly improve and interrogate the use of our methods. It is about transparency in the method and its improvement. If others can understand our approach and experiment with it, then they too can better adapt and refine it.

Aside from the two bioengineering horizon scans, we also used structured expert elicitation to identify promising research questions in biosecurity in collaboration with the Biosecurity Research Initiative at St Catharine's College. It was an attempt to craft a research agenda for the 2018 UK Biological Security Strategy (HM Government, 2018), rather than a set of priority future issues. We drew on a group of 41 experts from government, industry, and academia, who in turn reached out to an additional 168 experts. This expert pool put forward 450 policy-relevant questions that they believed would have the largest impact on UK biosecurity. We then facilitated a transparent, replicable process that drew from the IDEA protocol format: the questions were anonymously scored by participants, and a shortened list was forwarded for group discussion, followed by rescoring. The outcome was a list of the 80 most pressing, impactful, and unanswered questions for the UK Biological Security Strategy (Kemp *et al.*, 2021). Answering or funding research into these questions would aid the management of extreme biological risks.

Many of our horizon-scanning exercises have already proved prescient. The 2017 bioengineering horizon scan identified several short-term issues that have become more prominent since publication (Wintle *et al.*, 2017). For instance, the use of 'platform technologies for disease solutions' was a priority issue which discussed the use of biological platforms to streamline the production of vaccines and therapeutics. During the COVID-19 pandemic, these became widely used with numerous vaccine candidates making use of platforms used for non-Coronavirus diseases such as Influenza and Ebola. Issues such as military funding in bioengineering and human germline genome editing also had similar signals of accuracy. This is not a systematic review of the efficacy of the 2017 review, and it is too soon to conduct one. Nonetheless, it is a positive sign that the approach has value. Many of the questions identified in the 80 Questions exercise gained renewed prominence during the COVID-19 pandemic. These include questions around disease syndromic surveillance, ensuring quarantine compliance, and identifying and preventing unknown emerging infectious diseases.

Foresight is not just about understanding the future but also about bringing it into current thinking and policy. The results of the Managing Extreme Technological Risk Project have taken some encouraging steps on this front. For instance, the results of our 80 Questions exercise were presented to a House of Lords Committee on Risk Assessment and Risk Planning. Lalitha Sundaram also presented the results of the first bioengineering horizon scan to the Biological Weapons Convention Meeting of State Parties. The IDEA Protocol method used for the two bioengineering scans has already been partly adopted and used by the WHO under the foresight function of its Science Division. While promising, it remains to be seen whether these foresight activities will result in tangible changes in policy and action.

4.3 The Existential Risk Research Assessment

Anticipating extreme technological risk is a daunting endeavour. It is not just an exercise of identifying developments through horizon scans. It is also a matter of keeping track of the burgeoning research on global catastrophic and existential risks. Currently, there exists a significant gulf between the amount of published literature and the research that has been systematically reviewed, analysed, and summarised for decision-making. This is the 'synthesis gap'. It is particularly relevant in extreme technological and global catastrophic risks due to the high stakes involved, as well as the groundswell of literature in the past few decades.

Systematic reviews can be a time-consuming and onerous process. They require screening the titles and abstracts of potentially thousands of articles against a set of inclusion criteria. In this segment of the Managing Extreme Technological Risk Project, led by Gorm Shackelford, we combined crowdsourced reviews with machine learning to help address the synthesis gap. The use of machine learning text mining can reduce the human workload of a systematic review by 30–70%. The results were accumulated into an open-access database called the Existential Risk Research Assessment (TERRA), an inclusive, robust, and replicable method of producing a field-wide bibliography of global catastrophic risks. TERRA continues to be reviewed and updated on a monthly basis. The initial screening over 10 months involved 51 reviewers and covered 10,001 articles (Shackelford *et al.*, 2020).

Systematic literature reviews are one form of foresight and horizon scanning. They are already used in a wide array of disciplines including biomedicine (Hines *et al.*, 2019). TERRA provides the first steps towards being able to bring such tools to the task of anticipating extreme technological risk, and global catastrophic risks more broadly. The end goal of TERRA is to provide an automatically updated research tool which summarises the evidence on the different global threats and the effectiveness of different mitigation options. The current prototype of TERRA is one modest step towards this broader ambition.

4.4 Next Steps: A Systems Approach

Foresight for extreme technological risk ultimately requires a systematic approach: a way of tying together different tools from foresight, and forecasting, to provide a systemic view of plausible futures and the pathways that potentially lead to them. Each tool has its own strengths and weaknesses. Horizon scans provide a replicable and effective way of identifying issues on the margin of current thinking but tell us little of how these are interconnected and what policies are best suited to address them. Foresight alone tells us little of how to address extreme risks. For that, tools from risk analysis and robust decision-making under uncertainty are needed. In this section, I will sketch out how the tools developed in the METR project could be integrated together and combined with risk analysis to provide a comprehensive system for identifying and addressing extreme technological risk.

The IDEA protocol is a reliable and effective way for conducting horizon scans. The integration of the devil's advocates appears to be one promising way to better address black swan issues. However, there are still potential ways to improve the process. For forecasting, basic training to combat biases and practice heuristics for good judgement has been shown to significantly improve the Brier scores of novice participants (Chang *et al.*, 2016). Capturing and synthesising literature through machine learning platforms (such as the Meta platform for bioengineering (Meta, 2020) or TERRA (CSER, 2021)) could ensure that participants have a recent, common, and comprehensive view of relevant research. This improves the evidence base that experts can draw from.

Once issues have been identified through this enhanced Delphi technique, these could then be knitted together using systems mapping exercises. Using a process of structured expert elicitation, a group of scholars and practitioners could look to draw the priority issues into a single system via a causal loop diagram. This process is commonly referred to as collaborative conceptual mapping or modelling (Newell and Proust, 2012) and has many similarities to fuzzy cognitive maps (Jetter and Kock, 2014). This would overcome a key limitation of previous uses of the IDEA protocol by uncovering the interconnections between different issues and any common drivers. Scenarios and pathways towards them could then be used as a basis for understanding the route to extreme risks, early warning signs, and leverage points to prevent risks. A similar approach has already been suggested by Cremer and Whittlestone (2020) for locating 'artificial canaries' (echoing the use of canaries in coal mines) for AGI. This also echoes calls by Tetlock and Scoblic to identify clusters of short-term forecasting questions which can help inform us as to whether we are on the road to a ruinous scenario (Scoblic and Tetlock, 2020; Tetlock and Scoblic, 2021).

Tools for robust decision-making under uncertainty, such as the minimax principle (ranking actions by the use of such risk management tools has already been recommended for global risks such as climate change (Kunreuther *et al.*, 2013), their plausible worst-case scenarios), could then be applied to craft appropriate policies. Modified methods are perhaps more promising, such as the approach suggested by the Dasgupta Review on Biodiversity of ranking decisions by the weighted sum of their best- and worst-case outcomes (with the weighting reflecting risk adversity). The original term for this was the α-maxmin rule (Arrow and Hurwicz, 1977), but let's go with minimax+, which sounds more intuitive (Dasgupta, 2021).

An even more apt approach could be to feed in the foresight process in to citizen juries or assemblies, in essence, using a deliberative democracy of experts to inform a deliberative democracy of citizens. Such deliberative democratic methods have been shown to usually be superior to other decision-making methods in theory (Landemore, 2012) and practice (Landemore, 2020). Such deliberative experiments tend to give robustly good advice and also engender greater trust from the public, at the very least towards those involved with the deliberation (Boulianne, 2018). In this case, the panel or assembly, or online deliberative platform, could be asked to make use of the minimax+ principle to guide decision-making on catastrophic risks.

The suggested approach here is a synthetic and democratic foresight approach which looks to identify emerging risks, place them into a wider system, locate leverage points for change, and incorporate these into fair, open, and inclusive governance. The end outcome is a fusion of the best of democratic and expert decision-making methods to guide the world through complex catastrophic risks. We will also need to find the right balance between foresight and action. Often, resources may be better spent on resilience building and taking robust actions on Global Catastrophic Risk (GCR), rather than improving the calibration of forecasters or identifying yet another scenario. Indeed, at this stage, the former may be the far wiser way forward.

4.5 Conclusions

Managing extreme technological risk requires anticipation. Our project took the first steps towards providing a systematic way of conducting foresight on extreme risks from emerging technologies. Our approach intentionally focused on foresight rather than forecasting. Many forecasting tools have proved effective and dependable in making good probabilistic judgements about near-term, well-bounded events. This is not the domain in which extreme risks lurk. Rather than running tournaments and seeking higher Brier scores, we concentrated on understanding and prioritising plausible developments.

Our approach to foresight has concentrated on using horizon scanning, particularly through the IDEA protocol, to identify emerging issues in bioengineering and biosecurity. Our bioengineering scans in both 2017 and 2020, and our horizon scan of dual-use research of concern with the WHO in 2021, demonstrated how deliberative democracy among experts can discern critical risks and opportunities in the life sciences. The 2021 *80 Questions for UK Biosecurity* paper showcased how this expert deliberation can be fruitfully applied to identify research questions rather than emerging issues.

These horizon scans were complemented by our development of the TERRA platform: a crowd-sourced, machine learning algorithm for synthesising catastrophic-risk-relevant literature. Together, these tools provide the foundations for aggregating and reviewing the evidence for catastrophic technological risk, as well as spotting and prioritising threats on the margins of thinking that may be underrepresented in the literature.

Despite our apparent initial successes, foresight for extreme technological risk still faces several challenges. Extreme risks are usually black swans. Hunting these through foresight exercises is notoriously difficult. While our 2020 bioengineering scan took some first steps to better address the possibility of black swans, it is unclear how successful these were. This leads to the second difficulty. Given that only a few short years have elapsed, it is challenging to know how successful our methods have been. This is of course a hurdle for foresight more broadly.

The next steps are numerous. One is better synthesising foresight tools. This should include using systematic literature reviews such as those provided by TERRA to better inform expert elicitation and systems mapping exercises to interconnect issues identified in horizon scans and locate leverage points. This could include secondary exercises to identify early warning signals for priority issues identified in previous scans. Perhaps the most pressing and precious hurdle is bridging the schism between knowledge and action. That is, finding ways to weave the outcome of foresight research into deliberative democratic governance. We need to make democracy not just open but also anticipatory.

Yet, the future will always be shrouded in a fog of uncertainty. We need not only foresight but also the bravery to know its limits, to build systems that are anti-fragile and to be precautious with the dangerous development of Promethean gifts.

References

Amanatidou, E., *et al.* (2012). On concepts and methods in horizon scanning: Lessons from initiating policy dialogues on emerging issues. *Sci. Public Policy*, 28, 208–212.

Amer, M., Daim, T. U., and Jetter, A. (2013). A review of scenario planning. *Futures*, 46, 23–40.

Arrow, K. J. and Hurwicz, L. (1977). *Studies in Resource Allocation Processes*. Cambridge, UK: Cambridge University Press.

Barons, M. J., *et al.* (2018). Assessment of the response of pollinator abundance to environmental pressures using structured expert elicitation. *J. Apic. Res.*, 57, 593–604.

Boulianne, S. (2018). Building faith in democracy: Deliberative events, political trust and efficacy. *Polit. Stud.*, 67, 4–30.

Chang, W., Chen, E., Mellers, B., and Tetlock, P. (2016). Developing expert political judgement: The impact of training and practice on judgemental accuracy in geopolitical forecasting tournaments. *Judgm. Decis. Mak.*, 11, 509–526.

Cremer, C. Z. (2021). Deep limitations? Examining expert disagreement over deep learning. *Prog. Artif. Intell*. doi: 10.1007/s13748-021-00239-1.

Cremer, C. Z. and Whittlestone, J. (2020). Artificial canaries: Early warning signs for anticipatory and democratic governance of AI. *Int. J. Interact. Multimed. Artif. Intell.*, 6, 100–109.

CSER. (2021). The Existential Risk Research Assessment (TERRA). Available at: https://terra.cser.ac.uk (Accessed 29 January 2024).

Cuhls, K. (2003). From forecasting to foresight processes — New participative foresight activities in Germany. *J. Forecast.*, 22, 93–111.

Dasgupta, P. (2021). *The Economics of Biodiversity: The Dasgupta Review*. London: HM Treasury.

Flachsland, C., Marschinski, R., and Edenhofer, O. (2009). To link or not to link: Benefits and disadvantages of linking cap-and-trade systems. *Clim. Policy*, 9, 358–372.

Goodwin, P. and Wright, G. (2010). The limits of forecasting methods in anticipating rare events. *Technol. Forecast. Soc. Change*, 77, 355–368.

Grace, K., *et al.* (2018). When will AI exceed human performances? Evidence from AI experts. *J. Artif. Intell. Res.*, 62, 729–754.

Hanea, A. M., *et al.* (2017). Investigate discuss estimate aggregate for structured expert judgement. *Int. J. Forecast.*, 33, 267–279.

Hanea, A. M., *et al.* (2018). Classical meets modern in the IDEA protocol for structured expert judgement. *J. Risk Res.*, 21, 417–433.

Hemming, V., *et al.* (2018a). Eliciting improved quantitative judgements using the IDEA protocol: A case study in natural resource management. *PLoS One*, 13, 1–34.

Hemming, V., *et al.* (2018b). A practical guide to structured expert elicitation using the IDEA protocol. *Methods Ecol. Evol.* doi: 10.1111/2041-210X.12857.

Hines, P., *et al.* (2019). Scanning the horizon: A systematic literature review of methodologies. *BMJ Open*, 9, 1–9.

HM Government. (2018). *The UK Biological Security Strategy*. London: The Home Office.

Jetter, A. J. and Kok, K. (2014). Fuzzy cognitive maps for futures studies — A methodological assessment of concepts and methods. *Futures*, 61, 45–57.

Karger, E., Atanasov, P. D., and Tetlock, P. (2022). Improving judgements of existential risk: Better forecasts, questions, explanations, policies. *SSRN Electron. J.* Available at SSRN: https://ssrn.com/abstract=4001628 or http://dx.doi.org/10.2139/ssrn.4001628.

Kemp, L., *et al.* (2020). Bioengineering horizon scan 2020. *Elife*, 9, e54489.

Kemp, L., *et al.* (2021). 80 questions for UK biological security. *PLoS One*. https://doi.org/10.1371/journal.pone.0241190.

Kennicutt, M. C., *et al.* (2015). A roadmap for Antarctic and Southern Ocean science for the next two decades and beyond. *Antarct. Sci.*, 27, 3–18.

Kennicutt, M. C., *et al.* (2019). Sustained Antarctic research: A 21st century imperative. *One Earth*. https://doi.org/10.1016/j.oneear.2019.08.014.

Könnölä, T., *et al.* (2012). Facing the future: Scanning, synthesizing and sense-making in horizon scanning. *Sci. Public Policy*, 39, 222–231.

Kunreuther, H., Michel-Kerjan, E., and Tonn, G. (2016). *Insurance, Economic Incentives and Other Policy Tools for Strengthening Critical Infrastructure Resilience: 20 Proposals for Action*. Center for Risk Management and Decision Processes, The Wharton School, University of Pennsylvania. https://riskcenter.wharton.upenn.edu/wp-content/uploads/2020/12/WP2016 Dec_CIRI-Phase-I.pdf.

Kunreuther, H., *et al.* (2013). Risk management and climate change. *Nat. Clim. Chang.*, 3, 447–450.

Landemore, H. (2012). *Democratic Reason: Politics, Collective Intelligence, and the Rule of the Many*. Princeton, US: Princeton University Press.

Landemore, H. (2020). *Open Democracy: Reinventing Popular Rule for the Twenty-First Century*. Princeton, US: Princeton University Press.

Martin, B. R. (2010). The origins of the concept of 'foresight' in science and technology: An insider's perspective. *Technol. Forecast. Soc. Change*, 77, 1438–1447.

Mellers, B., Tetlock, P., and Arkes, H. R. (2019). Forecasting tournaments, epistemic humility and attitude depolarization. *Cognition*, 188, 19–26.

Mellers, B., *et al.* (2014). Psychological strategies for winning a geopolitical forecasting tournament. *Psychol. Sci.*, 25, 1106–1115.

Meta. (2020). Meta. Available at: https://scolarly.com/tools/menta (Accessed 29 January 2024).

Metaculus. (2021). Metaculus. Available at: https://www.metaculus.com/questions/ (Accessed 13 September 2021).

Neumayer, E., Plümper, T., and Barthel, F. (2014). The political economy of natural disaster damage. *Glob. Environ. Change*, 24, 8–19.

Newell, B. and Proust, K. (2012). Introduction to collaborative conceptual modelling. Working Paper: ANU Open Access Research. http://hdl.handle.net/1885/9386.

Owen, R. *et al.* (2012). Responsible research and innovation: From science in society to science for society, with society. *Sci. Public Policy*, 39, 751–760.

Parente, R. and Anderson-Parente, J. (2011). A case study of long-term Delphi accuracy. *Technol. Forecast. Soc. Change*, 78, 1705–1711.

Schultz, W. L. (2006). The cultural contradictions of managing change: Using horizon scanning in and evidence-based policy content. *Foresight*, 8, 3–12.

Scoblic, P. J. and Tetlock, P. E. (2020). Better crystal ball: The right way to think. *Foreign Aff.*, 99, 10–19.

Shackelford, G. E., *et al.* (2020). Accumulating evidence using crowdsourcing and machine learning: A living bibliography about existential risk and global catastrophic risk. *Futures*, 116, 102508. https://doi.org/10.1016/j.futures.2019.102508, https://www.sciencedirect.com/science/article/pii/S0016328719303702.

Stilgoe, J., *et al.* (2013). Developing a framework for responsible innovation. *Res. Policy*, 42, 1568–1580.

Sutherland, W. J. and Woodroof, H. J. (2009). The need for environmental horizon scanning. *Trends Ecol. Evol.*, 24, 523–527.

Sutherland, W. J., *et al.* (2019). Ten years on: A review of the first global conservation horizon scan. *Trends Ecol. Evol.*, 34, 139–153.

Taleb, N. N. (2007a). *The Black Swan*. Harlow, England: Penguin Books.

Taleb, N. N. (2007b). *Fooled by Randomness: The Hidden Role of Chance in Life and in the Markets*. Harlow, England: Penguin Books.

Taleb, N. N. (2012). *Antifragile: Things That Gain from Disorder*. Harlow, England: Penguin Books.

Taleb, N. N. (2020). On the statistical differences between binary forecasts and real world payoffs. *Int. J. Forecast.*, 36(4), 1–13.

Taleb, N. N. and Tetlock, P. E. (2013). On the difference between binary prediction and true exposure with implications for forecasting tournaments and decision making research. *SSRN Electron. J.* doi: 10.2139/ssrn.2284964.

Taleb, N. N., *et al.* (2020). On single point forecasts for fat-tailed variables. *Int. J. Forecast.* doi: 10/1016/j.ijforecast.2020.08.008.

Tetlock, P. and Gardner, D. (2016). *Superforecasting: The Art and Science of Prediction.* London: Random House.

Tetlock, P. E. and Scoblic, P. J. (2021). Can we reconcile narrativist and probabilistic modes of thinking? *Foresight,* (63), 21–25.

Tetlock, P. E., *et al.* (2022). False dichotomy alert: Improving subjective-probability estimates vs. raising awareness of systemic risk. *Int. J. Forecast.* doi: 10.1016/j.ijforecast.2022.02.008.

Wintle, B. C., *et al.* (2017). A transatlantic perspective on 20 emerging issues in biological engineering. *Elife,* 6, 61–21.

Chapter 5

Evaluating Extreme Technological Risk: A Social Contract Based Approach

S. J. Beard and Patrick Kaczmarek

5.1 Introduction

Extreme technological risk necessarily involves significant intergenerational considerations. On the one hand, the development of risky technologies, like biotechnology, nanotechnology, and artificial intelligence (AI), is often justified by their ability to alter the trajectory of humanity's future towards what is maximally beneficial and safe for us. On the other hand, these technologies carry the potential to cause global catastrophes, such as human extinction, the collapse of human civilisation, or a permanent shift in humanity's future trajectory that would be harmful and/or dangerous. While these changes would most likely have significant impact on the present generation, that is only a small part of their effects and we cannot truly evaluate them without taking the intergenerational perspective (Parfit, 1984).

The standard means for undertaking such an evaluation is by impartially weighing up costs and benefits from a consequentialist moral standpoint, such as utilitarianism. This has been, implicitly or explicitly, the position taken by the majority of researchers in the field of Existential Risk Studies, and while there are undoubtedly some important philosophical considerations that need to be addressed when carrying out such evaluations, the results have been generally serviceable and well founded (Millett and Snyder-Beattie, 2017; Wolff, 2021). Nevertheless, it would be bad if this was the only way in which we could capture the moral importance of these aspects of extreme technological risk. For one thing, consequentialist approaches to moral theory in general, and cost–benefit analysis in particular, are controversial and many people do not accept them. While there is general academic acceptance of these views across, say, most leading economics departments, many anglophone philosophy departments, much of contemporary 'start-up culture', and regulatory institutions in the UK and USA, there is profound scepticism and even hostility towards them among most leading public health departments, many non-anglophone philosophy departments, most of contemporary 'activist culture', and regulatory institutions in continental Europe. Another problem is that each of us should be humble enough to recognise that an attitude of moral uncertainty is epistemically appropriate, and that even if we were to broadly agree with consequentialism (which the authors of this chapter do), it may still turn out to be incorrect and superseded by other moral theories that could be very different in both substance and spirit.

For these reasons, we consider it prudent for philosophers with an interest in extreme technological risk to constructively engage with other moral perspectives and develop evaluative approaches that are compatible with these. In service of that goal, we here summarise one such approach based upon the social contract tradition.

5.2 The Intergenerational Nature of the Social Contract

The social contract tradition holds, roughly, that the content of morality derives from agreement between all those in the moral domain (Ashford and Mulgan, 2007). If the 'social contract' were understood as an *actual* agreement made between various parties, and as such required that only those parties who could presently submit to, or negotiate, an agreement can be understood as parties to it, then this would seem to rule out any effects of our actions on merely possible future people as an irrelevant moral consideration.

It is true that the pioneers of this tradition, such as Thomas Hobbes (1651) and Jean-Jacques Rousseau (1762), did appear to conceive of the social contract as some form of actual agreement that people make, so as to either save themselves from anarchy (according to Hobbes) or trade their natural freedoms for a better quality of life (according to Rousseau).

However, from early on, people saw that even *if* this made sense at a moral level, it was clearly a misrepresentation of how we actually live our lives. As David Hume, writing about the social contract in 1752, argued:

> Were you to preach ... that political connections are founded altogether on voluntary consent or a mutual promise, the magistrate would soon imprison you ... if your friends did not before shut you up as delirious for advancing such absurdities. (Hume, 1753, 306)

Since then, nearly everyone who has engaged with social contract theory has claimed, not that this is an actual contract to which everyone consents, but rather a hypothetical contract that people either *would* agree to under certain conditions, *could* agree to without violating certain stipulations, or *ought* to agree to if they thought about it in a certain way. Understood in this sense, the possibility of a social contract that extends beyond those presently alive becomes much less problematic, and indeed such a contract was proposed by Edmund Burke as early as 1790.

'Society is indeed a contract', Burke maintains; however, it is different in both form and scope to the kinds of contract we are used to dealing with. While such '[s]ubordinate contracts for objects of mere occasional interest may be dissolved at pleasure', society 'is to be looked on with other reverence, because it is not a partnership in things subservient only to the gross animal existence of a temporary and perishable nature' (Burke, 1790, paragraph 165). Rather:

> It is a partnership in all science; a partnership in all art; a partnership in every virtue and in all perfection. As the ends of such a partnership cannot be obtained in many generations, it becomes a partnership not only between those who are living, but between those who are living, those who are dead, and those who are to be born. (*Ibid.*)

In other words, to be alive is not to be engaged in some solitary pursuit to meet your own present needs, but rather to engage in cooperative and perpetual projects for the betterment of all. We ourselves, the present generation, have benefited enormously from previous generations who have engaged in such projects, while future generations are relying upon us to do the same. What is more, this process of working for the good of posterity is not simply some happy accident of human nature (akin to Adam Smith's 'invisible hand'); rather, we choose to do this because it matters to us. As Samuel Scheffler (2010, p. 269) puts it, 'what would it mean to value things but, in general, to see no reason of any kind to sustain them or retain them or preserve them or extend them into the future?'

This notion of humanity as involved in some perpetual collective partnership to realise 'Burkean projects', such as science, art, virtue, and perfection, for the benefit of future generations is quite common amongst futurists, as is the notion that either causing or allowing a catastrophe to threaten these projects would betray the trust that previous generations have placed in us to complete what they began. For instance, Wendel Bell (1993) and Richard Slaughter (1994) both support the notion that present generations have an obligation to see that humanity's 'important business' is not left unfinished. Bruce Tonn, combining the work of these two with his own Rawlsian analysis and some basic assumptions about the future trajectory of humanity, argued that our obligations under such a partnership imply that 'the acceptable probability of human extinction for the imaginable future is 2×10^{-20}' (Tonn, 2009, p. 432). Nick Bostrom (2013, p23) has similarly claimed that 'we might also have custodial duties to preserve the inheritance of humanity passed on to us by our ancestors and convey it safely to our descendants. We do not want to be the failing link in the chain of generations'. Most recently, Leopold Aschenbrenner has developed a version of these ideas he calls Burkean Longtermism (2021).

Yet, such lines of reasoning have received little serious attention within analytic moral philosophy; however, there are clear parallels in other, more communitarian, traditions to intergenerational ethics including, but not limited to, those explored in contemporary African philosophy (Behrens, 2012). In what follows, we offer a preliminary analysis of how such a theory might be developed by summarising three key lines of argument concerning the nature of our duties to the past, the future, and humanity as a whole.[1] These arguments are developed to be compatible with a particular version of the social contract tradition associated with the work of Tim Scanlon (1998). While not necessarily reflecting all of social contract theory as a whole, we chose this form of contractualism for three reasons. Firstly, it is most commonly cited in contemporary analytical

---

[1] These are arguments that we have made elsewhere, and we are grateful to the editors of those journals, *Utilitas* and *Argumenta*, for permission to reprint them here. Respectively, see Kaczmarek and Beard (2020) and Beard and Kaczmarek (2019).

moral philosophy. Secondly, it is the form of contractualism with which what little literature has been produced on these issues tends to engage. Thirdly, it allows for a greater degree of formalisation than most versions of contractualism, and while we offer only the bare minimum of formality in this chapter, we hope eventually to develop a theory that can be put to work on technical issues requiring such formality, such as risk assessment and population ethics.

5.3 Our Obligations to the Past

Consider:

> Through no fault or choice of his own, Jeff is very sick. He desperately needs a liver transplant. Although he is not obliged to do so, a stranger called Michael gives Jeff part of his liver at the cost of reducing his own lifespan by ten years. After the procedure, Jeff chooses to drink heavily, and he dies from cirrhosis shortly afterwards.

In considering a case like this, it is our view that Michael does the right thing by Jeff in donating part of his liver but that, by being reckless, Jeff does the wrong thing, not only by himself but also by Michael, and this makes the state of affairs worse than it would otherwise have been.

Let's start by distinguishing two features of this case that contribute to the moral status of Michael's sacrifice in donating part of his liver to Jeff. Firstly, there is the question of whether Michael acted appropriately as a moral agent. The answer to this question doesn't depend on what Jeff does next. All that matters is whether Michael himself is sensitive to the moral reasons involved in this decision (including the expected benefit to Jeff at the time of transplant, the expected cost to Michael, and any further considerations, such as special duties Michael may have to Jeff). Secondly, we can ask whether other people in this case are sensitive to the moral reasons that flow from Michael's decision, and in particular the fact that, by deciding as he has, Michael has been harmed.

Despite the fact that it was morally justified, shortening Michael's life is a bad thing that, all else being equal, we would remove from the world if we could. Jeff's actions post-surgery appear to make this bad thing even worse, not because they make Michael's life any worse, but because they fail to realise the full extent of the benefits made possible by Michael's sacrifice. When choosing to donate part of his liver, Michael did the right thing because he judged that Jeff could expect to get much more than the ten years of additional life that Michael himself had sacrificed. He trusted Jeff to realise this full benefit from his sacrifice, making it, in other words, worthwhile. However, Jeff's actions post-surgery meant that he actually benefited by a lot less than Michael had lost. Jeff thus betrayed Michael's trust, and thereby wronged him.

What is true in this rather simple case also applies, we would argue, in cases at a much grander scale. Although our actions today can neither prevent nor mitigate the harms that were involved in past sacrifices, they can still do right or wrong by those who made those sacrifices by supporting or frustrating the Burkean projects they were engaged in, and thus making their losses more or less worthwhile. Our ancestors laboured hard under conditions we would find intolerable to secure benefits for their descendants, fought to end slavery and promote human rights, participated in the creation of feats of engineering, artistry, science, and technological development whose benefits they would never see, and otherwise made numerous sacrifices for our sake. The worthwhileness of such sacrifices could well alter the moral reasons we have for acting to varying degrees and, when contravened in the absence of adequate justification, can indicate a lack of concern for the interests of others. Let us call this the Moral Worthwhileness Intuition (hereafter abbreviated 'MWI').

To accept MWI is to say that we have a *pro tanto* reason to promote the interest of future generations as Burke suggested, given the many sacrifices that our forebears have made for their sake. Even if it would not wrong any future people if ours was to be the last generation of humans, it still seems to wrong those who had anticipated and made grave sacrifices for the benefit of future generations and the long-term future of our species. Only if future generations actually exist, and enjoy the fruits of these Burkean projects, will those sacrifices be worthwhile.

5.3.1 *The value of worthwhileness*

Yet, merely accepting MWI still leaves important questions to be answered about the nature of our obligations to the past, and whether these give us sufficiently strong reason to both work for the benefit of posterity and to prevent catastrophes that would destroy humanity's potential.

One can appreciate the general form that such a theory might take by considering the fact that one who has made a sacrifice would reasonably object to another person acting in such a way as to make their sacrifice worthless. This is because making this sacrifice was morally supererogatory and done purely because of the benefits it would bring to others. Actually realising these benefits, however, was not fully in the gift of the sacrificer and is thus left as *unfinished business* for others to complete: including both the person benefited and potentially other intermediaries between the sacrificer and the beneficiary. This unfinished business takes the form of additional obligations to act in ways that might otherwise have been beyond the call of duty.

To put this more formally, on the *Unfinished Business Account*:

All else being equal, if actor p reasonably judges that performing a supererogatory act ϕ at great sacrifice to herself will enable beneficiary q to achieve a greater good, then failure to promote the good made possible by ϕ wrongs p, *even if p is no longer alive.*

Of course, this does not imply that this wrongdoing is always decisive, or even significant, in our moral deliberations. For instance, common sense tells us that I cannot enslave another person simply by accepting some mild sacrifice whose benefits would depend greatly on what she then chose to do. At the limit, it is clearly not possible to oblige another person to accept a greater sacrifice than that which one originally accepted or to accept a sacrifice that, when combined with our own, would yield a total cost that was greater than the total benefit we could together have achieved. While Michael could reasonably expect Jeff to refrain from drinking himself to death as a result of sacrificing part of his liver, he could not expect Jeff to undergo torture, even if this would ultimately produce some even greater good.

More substantially, it seems only reasonable that if our obligations to the past are to play a decisive role in our moral choices, then they should be at least broadly in line with our long-term interests and consistent with our conception of the good life, even if the required action is not what we ourselves would have chosen. For instance, we can easily imagine that Jeff looking after his liver is consistent with his own conception of the good life, even if he would actually much rather stare down the bottom of a bottle. However, were we to require him to convert to Catholicism and eat nothing but Ramen noodles, this may well be inconsistent with his conception of the good life. At the grander scale, this implies that, while we may be bound to complete projects of genuinely shared human value and to promote the general interests of future people, we have no obligation to realise the mere whims or peculiar interests of our forebears concerning the future.

One may wonder whether we really can have obligations to those who have made sacrifices even once they are dead. There is nothing either new or radical in this view (see, e.g., Marquis, 1985); however, there are some philosophers who claim that this is not possible because, for instance, 'an act cannot be wrong if it would be worse for no one' (Parfit, 2017, p. 118). We can never wrong our forebears, according to such views, because nothing we now do can make them better or worse off — after all, they are dead. However, it is unclear just how substantial this objection proves, given that, even if we don't think they are directly harmed, most of us do put great store on respecting the past and those who lived in it.

Nevertheless, this deserves some exploration. Here, we assume the truth of Carl Wellman's notion of *surviving duties*; as he writes, '[d]eath destroys a person's capacity for agency and thereby destroys an essential qualification for being a moral right-holder. At the same time, death sometimes leaves the moral reasons intact so that, together with the qualifications of the surviving duty-bearer, these can ground a surviving duty' (Wellman, 1995, pp. 156–157). Our claim is that, so far as we have a duty to realise the value of others' sacrifices to realise the grand projects of humanity, these duties do not perish with the death of the person who performed this sacrifice; nor does their strength depreciate.[2]

5.4 Our Obligations to the Future

Still, as Burke says, our obligations under the social contract not only extend to those who are living and those who are dead but also to 'those who are to be born'. Here, the potential for our choices to make things worse is not in doubt. However, there are still those who object that, from the point of view of the social contract, these negative impacts are not morally salient.

---

[2]For a selection of objections to this argument and our responses to them, see Kaczmarek and Beard (2020).

The most common defence of this claim stems from the Scanlonian version of contractualism according to which, 'an act is wrong if its performance under the circumstances would be disallowed by any set of principles for the general regulation of behaviour that no one could reasonably reject for informed, unforced, general agreement' (Scanlon, 1998, p. 153). This is to say, a principle is morally sound just in case it is justifiable to each person, where such justification is attentive to the interests, complaints, and other claims of the individuals in question.

To define this view more precisely, Scanlon places two restrictions on the reasons that can be offered for rejecting a moral principle. Firstly, there is what Derek Parfit has called the *Impersonalist Restriction:* 'Impersonal values are not themselves grounds for reasonable rejection' (*ibid.*, p. 222; cf. Parfit, 2011, p. 214).

This means that if it is wrong to fail to bring future people into existence, then this must be because of the value of this existence to these people. It cannot simply be a recognition that these people's lives would be good from some third-party point of view — as Henry Sidgwick memorably put it, 'the point of view of the universe' (Sidgwick, 1907, p. 382).

Moreover, there is what Parfit dubs *The Individualist Restriction:* 'The justifiability of a moral principle depends only on individuals' reasons for objecting to that principle and alternatives to it' (Scanlon, 1998, p. 229; cf. Parfit, 2011, p. 193). This means that the value of future people's lives to them must rest in the fact that this promotes some interest that these people actually have, and that we fail to take account of the people's interest in these outcomes, not the outcomes themselves.

5.4.1 *Finneron-Burns's argument*

In a recent paper, Elizabeth Finneron-Burns (2017) has developed a series of arguments based on these restrictions, which aim to show that the potential loss of future generations, including from possible technological catastrophes, should not (that is, non-derivatively) factor into our choices. In particular, she argues that it would be morally irrelevant that our actions might lead to many people never coming into existence because there would be no one *there* who has an interest in being brought into existence, and whom we failed to attend to. To be sure, she then goes on to argue that causing such a disaster would still be wrong, but only due to its impact on those already living.

Finneron-Burns states her principal argument against the need to take account of the impact of catastrophes on merely possible people as follows:

> [W]e can only wrong someone who did, does or will actually exist because wronging involves failing to take a person's interests into account. When considering the permissibility of a principle allowing us not to create Person X, we cannot take X's interest in being created into account because X will not exist if we follow the principle. By considering the standpoint of a person in our deliberations, we consider the burdens they will have to bear as a result of the principle. In this case, there is no one who will bear any burdens since if the principle is followed (that is if we do not create X), X will not exist to bear any burdens. So, only people who do/will actually exist can bear the brunt of a principle, and therefore occupy a standpoint that is owed justification. (*Ibid.*, p. 331)

According to this argument, we can only wrong people by failing to take their interests into account (i.e., the individualist restriction), and people's interests only extend to not being burdened. Since merely possible people do not have interests, as people who do not exist do not bear any burden by virtue of their non-existence, we, therefore, cannot wrong them.

5.4.2 *How we might still have obligations to future people*

One reply to this argument comes from Rahul Kumar (2018). Kumar agrees with Finneron-Burns that we do not *harm* future people by causing them not to exist. However, he argues that a contractualist can still object to causing or allowing catastrophes that would cause future people not to exist, even though this does not constitute a harm to them. There are cases, he argues, in which we act wrongly by failing to take a person's interests into account even though we do not, in fact, harm or burden them in any way. One example in which we can be said to do this is when we act negligently, or even viciously, towards a person but fail to do any harm through sheer dumb luck. If this is possible, the argument goes, then why is it not equally possible for us to fail to take somebody's interests into account in cases in which that person is not burdened because we cause them not to exist?

In an earlier paper, Finneron-Burns (2016) objects to this line of argument on the grounds that, while it is possible to wrong a person without harming them, this can only be achieved if there is a relevant 'standpoint', or cluster of personal characteristics and interests that an individual can occupy, whose interests are not being taken into account. There are, she admits, a great variety of such standpoints that may be morally salient. Any of these could give us reasons for rejecting a moral principle on the grounds that it does wrong by a person who might occupy this standpoint, even though it does not actually harm anyone because nobody, in the end, actually occupies it. However, she goes on to claim that this approach cannot be applied to the interests of merely possible people because standpoints can only describe positions that could be occupied by actual people and existence is not an interest that can be attached to a standpoint (*ibid.*, p. 1156ff).

According to another reply, it is claimed that we fail to take an individual's interests into account in cases when they do not exist because non-existence is equivalent in value to a person existing with a neutral life — that is, a life that is neither good nor bad for them — so that they would be genuinely better off if they had received a good life instead. On this account, we can be said to burden merely possible future people because we have brought about a state, non-existence, that is worse for them than a possible happy life, which they might have had if we had acted otherwise (Fleurbaey and Voorhoeve, 2015; Arrhenius and Rabinowicz, 2015; cf. Pummer, 2019). This makes non-existence a real burden compared with the possibility of existence with a good life, and if we choose this option, then we can be said to harm in the same way as if we had simply made the life of an actual person much worse so that it were no longer good for them.

However, from a Scanlonian perspective, this raises the troubling prospect of ethical paralysis. Since everybody who does not come into existence with a life worth living is, on this account, harmed in just the same way as a person forced to live a neutral life, then it seems that all non-existing people have strong reasons to reject any principle that does not bring them into existence (cf. Kavka, 1975). Since all, or virtually all, principles may have some effect upon the identity of those who exist in the future, this would mean that there is always someone whose interests we are failing to take account of when applying such principles, and hence no, or almost no, ethical principles will be left on the table. As Finneron-Burns puts it, 'even basic, everyday activities would be morally wrong since they prevented certain people from coming to exist' (Finneron-Burns, 2016, p. 1157).

We believe that there is another, better, reply to Finneron-Burns's argument that explains why we should care about the loss of future generations. According to this response, we do wrong by a merely possible person whom we denied existence, not because we bring about a state of affairs that has zero value for them, but because we do not bring about a state of affairs that would have had a positive value for them when we could have done so (without sacrificing anything of comparable value).

When we make choices that determine whether a possible future person will come into existence with a life worth living, we choose between an outcome that will be good for them and one that will not be good for them, even though, since they do not exist in this second outcome, it will not be worse for them either. To put it another way, on this proposed view, we can be said to fail to take a person's interests into account because we have failed to benefit them, even though we have not burdened them either. This kind of argument was first proposed by Krister Bykvist (2007). However, we focus here on the version offered in a posthumously published paper by Derek Parfit (2017).

Parfit labels the kind of position presented by Finneron-Burns the 'No Complaints Claim', which he neatly summarised by the slogan '[a]n act cannot be wrong unless there is or will be someone whom this act has wronged' (*ibid.*, p. 136). He admits that this view has considerable psychological and moral force; however, Parfit is keen to point out that it only describes part of morality. In particular, it rests upon the idea that what we owe each other is exhausted by the reasons people give us not to act badly towards them, what Parfit refers to as 'reasons of non-maleficence'. In Scanlon's terms, these are the reasons that a person might have for rejecting the principle on which we are acting.

In some of his earlier writings, Parfit appeared to accept Scanlon's line of argument, going so far as to say that the wrongness of denying future people the benefit of existence cannot stem from the personal reasons that these people themselves would give us to bring them into existence (Parfit, 2011, p. 8). However, in what would be his final paper, Parfit moved beyond this position and argued that we could have personal reasons to bring people into existence. These would be reasons of benevolence to do what is good for a person and promote their 'personal goodness'. For him, the difference between these reasons and reasons of 'non-maleficence' seemed, at most, a matter of degree rather than of type, so that if a moral theory holds that we should respect reasons of non-maleficence, it should not also claim that reasons of benevolence have no moral weight whatsoever (Parfit, 2017, p. 136ff).

While he discussed this possibility, Parfit did not offer any firm conclusions about it. However, his line of argument appears to us to suggest that Finneron-Burns's argument is far weaker than she claims. Specifically, things are hardly so cut and dry for the Scanlonian contractualist. They may (and if Parfit is right, should) accept a 'Wide Person-Affecting Principle', according to which:

> One of two outcomes would be in one way worse if this outcome would be less good for people, by benefiting people less than the other outcome would have benefited people. (*Ibid.*, p. 129)

Unlike the No Complaints Claim, Parfit's proposed principle does not restrict us to reasons of non-maleficence and instead considers the whole of morality, including how we might benefit people. Of course, there is no great difficulty in saying that a person whom we might bring into existence and who would have a good life could give us reasons to choose to bring them into existence, on the grounds of beneficence, because they would be glad to be alive in much the same way that a person whom we might bring into an awful, wretched existence provides in us reason to choose not to bring them into existence, on the grounds of non-maleficence, because they would regret our actions. To put this more formally:

> If person *p* is caused to exist and to have a life that is worth living, then that is good for this person, giving him or her an 'existential benefit'; all else being equal, denying this benefit is wrong, even though *p* is not made worse off as a result.

Finneron-Burns claims that 'wronging involves failing to take someone's interests into account' (2017, p. 5); however, when arguing that we cannot wrong future generations, she equates a person's interests with the avoidance of bearing burdens. Yet, many of us would say that avoiding burdens constitutes an incomplete — nae, impoverished — picture, and that our interests can equally be served by receiving benefits. When we fail to bring somebody into existence, we can legitimately be said to have failed to take their interests into account, even if we didn't harm them, because we also didn't benefit them (by bringing them into existence) when we might easily have done so. It is in this respect that the interests of possible people count as reasons in favour of avoiding the destruction of the future, and not, as Finneron-Burns contends it would have to be, that we should militate against such threats only to the extent that doing so avoids harm. Since one of the outcomes that we might choose is good for these people, and hence serves their interests, failing to bring this outcome about would fail to take account of their interests and thus constitute a kind of mean-spiritedness, even though the actual outcome is not worse for them. Hence, we can be said to have done wrong by failing to act in the way that would have been good for these people, and thus denied them this existential benefit, whether or not the interests we fail to take into account can be said to actually exist in the outcome that we eventually bring about.

Of course, Finneron-Burns and Scanlon have their reasons for preferring to focus on moral burdens. One of these is surely that their brand of morality, in general, has a negative tone, since the aim of Scanlonian contractualism is to identify which principles are morally impermissible by working out which ones people could reasonably reject (a formulation that traces back to Immanuel Kant rather than the social contract tradition). Parfit himself has done much to criticise the motivation for this stern view, and to argue that it must be counterbalanced by a moral philosophy that considers which principles are required by virtue of being the ones that people could 'rationally accept.' However, he also believed that these two approaches to morality ultimately arrive at the same result — or at least that, as he once put it, that '[they] are climbing the same mountain on different sides' (Parfit, 2011, p. 373).

However, we do not think that the point we are concerned with here requires us to accept this further claim of Parfit's. Perhaps the theoretical equivalence between impermissibility, reasonable rejection, and *burdening* is of symbolic importance. But there is no reason why it would not be of equal moral importance to equate impermissibility, reasonable rejection, and *not benefiting*. Indeed, for the kinds of so-called 'same population' cases with which Scanlonian contractualists are overwhelmingly concerned, the two would appear to amount to very much the same thing, and it is only when one considers how our actions might affect the size, or even existence, of future generations that we see any difference in their practical implications. To claim that in these latter cases one can only go along with Scanlon's original concept of moral burden, and that since nobody is burdened by their non-existence, causing them not to exist cannot be wrong appears to us to be going too far. Why do we not owe some justification to these possible people for failing to do what would have been good for them, and why would that not provide equal grounds for wrongness via reasonable rejection and hence to including their interests in our conception of the social contract?[3]

5.5 Our Obligations to Humanity as a Whole

Thus, it seems plausible to us that Burke had it right when he asked us to view the social contract as something that extends across generations, and that we can have obligations to both the past and the future that require us to seek to promote the interests of future people while preserving humanity from extreme technological risk. However, we may also have reasons that go beyond even this expansive contract.

There are many ways in which we can think about the value of the future. According to a distinction put forth by Christine Korsgaard, it is possible to both value something for its own sake, or as an end in itself, which is its 'final value', and to value it for the sake of something else, or as a means to an end, which is its 'instrumental value' (Korsgaard, 1983). Our arguments so far have taken the form of claiming that we should seek to protect the future of humanity because it is good for people, promoting the interests of future generations and making the sacrifices of past and present generations more worthwhile. However, some philosophers have also argued that humanity as a whole also possesses final value, and that this is something the social contract requires us to promote, if we can, as well.

Johann Frick in particular has done most to advance this idea. The argument, as Frick presents it (2017), begins with the observation that people commonly attribute final value to a range of phenomena, such as languages, species, and cultures, that they see as worthy of protection for their own sake. This suggests 'that humanity too, with its unique capacities for complex language use and rational thought, its sensitivity to moral reasons, its ability to produce and appreciate art, music, and scientific knowledge, its sense of history, and so on, should be deemed to possess final value' (*ibid.*, p. 359). And, heck, why not?

---

[3] Finneron-Burns provides further supplementary arguments, which we respond to in Beard and Kaczmarek (2019).

As we noted in the first section of this chapter, our intergenerational social contract exists not only because we happen to rely upon previous generations for our lives and livelihoods while future generations are reliant upon us for the same. It also exists because working for posterity and the betterment of humankind simply is something that matters to people (including ourselves) and that gives meaning and purpose to our lives. Having this collective Burkean project cut short is something that we should all, plausibly, be deeply concerned about for its own sake, and not simply because we owe it to the interests of others, especially if we were to discover that humanity alone possesses these capacities, sensitivities, and abilities. As Frick says, that 'would be very bad, indeed one of the worst things that could possibly happen' (*ibid.*, p. 344).

5.6 Conclusion

The idea of an intergenerational social contract encompassing not only presently existing people but also those who have come before and those who are yet to come has a long history and seems to be attractive to many who worry about extreme technological risk. It seems a shame therefore that it has received so little attention within analytical moral philosophy, and that what attention there has been is so often used to provide arguments against it. In this chapter, we have tried to address some of these objections and offer positive arguments for why our obligations to the past, the future, and humanity as a whole form important aspects of the social contract. In doing so, we hope to have provided a foundation for new ways of evaluating extreme technological risk that would appeal to communities who are opposed to utilitarianism, whether out of instinct, ideology, or conviction.

References

Arrhenius, G. and Rabinowicz, W. (2015). The value of existence. In I. Hirose and J. Olson, (eds.), *The Oxford Handbook of Value Theory*. Oxford: Oxford University Press, pp. 424–443.

Aschenbrenner, L. (2021). Burkean Longtermism. Personal blog dated 26th July 2021. Available at: https://www.forourposterity.com/burkean-longtermism/.

Ashford, E. and Mulgan, T. (2007). Contractualism. In E. N. Zalta (ed.), *The Stanford Encyclopedia of Philosophy* (Summer 2018 Edition). Available at: https://plato.stanford.edu/archives/sum2018/entries/contractualism/.

Beard, S. and Kaczmarek, P. (2019). On the wrongness of human extinction. *Argumenta*, 5(1), 85–97.

Behrens, K. G. (2012). Moral obligations towards future generations in African thought. *J. Global Ethics*, 8(2–3), 179–191.

Bell, W. (1993). Why should we care about future generations? In H. F. Didsbury, Jr., (ed.), *The Years Ahead*. Bethesda, MD: World Future Society, pp. 25–41.

Burke, E. (1790). *Reflections on the Revolution in France: And on the Proceedings in Certain Societies in London Relative to that Event: In a Letter Intended to Have Been Sent to a Gentleman in Paris*. London: J. Dodsley.

Bykvist, K. (2007). The benefits of coming into existence. *Philos. Stud.*, 135(3), 335–362.

Finneron-Burns, E. (2016). Contractualism and the non-identity problem. *Ethical Theory Moral Pract.*, 19, 1151–1163.

Finneron-Burns, E. (2017). What's wrong with human extinction? *Can. J. Philos.*, 47(2–3), 327–343.

Fleurbaey, M. and Voorhoeve, A. (2015). On the social and personal value of existence. In I. Hirose and A. Reisner (eds.), *Weighing and Reasoning: Themes from the Philosophy of John Broome*. Oxford: Oxford University Press, pp. 95–109.

Frick, J. (2017). On the survival of humanity. *Can. J. Philos.*, 47(2–3), 344–367.

Hume, D. (1753). Of the original contract. In *Essays and Treaties on Several Subjects*. Edinburgh: Kincaid and Donaldson.

Hobbes, T. (1651). *Leviathan or The Matter, Forme and Power of a Commonwealth Ecclesiasticall and Civil*. London: Andrew Crooke.

Kaczmarek, P. and Beard, S. (2020). Human extinction and our obligations to the past. *Utilitas*, 32(2), 199–208.

Kavka, G. (1975). Rawls on average and total utility. *Philos. Stud.*, 27(4), 237–253.

Korsgaard, C. M. (1983). Two distinctions in goodness. *Philos. Rev.*, 92(2), 169–195.

Kumar, R. (2018). Risking future generations. *Ethical Theory Moral Pract.*, 21(2), 245–257.

Marquis, D. (1985). Harming the dead. *Ethics*, 96(1), 159–161.

Millett, P. and Snyder-Beattie, A. (2017). Existential risk and cost-effective biosecurity. *Health Secur.*, 15(4), 373–383.

Parfit, D. (1984). *Reasons and Persons*. Oxford: Oxford University Press.

Parfit, D. (2011). *On What Matters* (Vol. 2). Oxford: Oxford University Press.

Parfit, D. (2017). Future people, the non-identity problem, and person-affecting principles. *Philos. Public Aff.*, 45(2), 118–157.

Pummer, T. (2019). The worseness of nonexistence. In E. Gamlund and C. Solberg (eds.), *Saving People from the Harm of Death* (pp. 215–228). New York: Oxford University Press.

Rousseau, J. J. (1762) *Du contrat social; ou Principes du droit politique.* Amsterdam: Marc Michel Rey.

Scanlon, T. M. (1998). *What We Owe to Each Other*. Cambridge: Harvard University Press.

Scheffler, S. (2010). *Equality and Tradition*. Oxford: Oxford University Press.

Sidgwick, H. (1907). *The Methods of Ethics* (7th edn.). London: MacMillan and Co.

Slaughter, R. A. (1994). Why we should care for future generations now. *Futures*, 26(10), 1077–1085.

Tonn, B. E. (2009). Obligations to future generations and acceptable risks of human extinction. *Futures*, 41(7), 427–435.

Wellman, C. (1995). *Real rights*. Oxford: Oxford University Press.

Wolff, J. (2021). Risk and the regulation of new technologies. In T. Matsuda, J. Wolff, and T. Yanagawa (eds.), *Risks and Regulation of New Technologies* (pp. 3–18). Singapore: Springer.

https://doi.org/10.1142/9781800614826_0006

Chapter 6

Responsibility and the Management of Extreme Technological Risk: Bio(techno)logical Risk

Catherine Rhodes

6.1 Introduction

The third strand of our METR research programme focused on responsible innovation and extreme technological risk, including an in-depth examination of lessons to be learnt from the Fukushima nuclear disaster, and engagement activities with policy, technology, and other academic communities around risks associated with emerging technologies, particularly in the areas of artificial intelligence (see Chapter 7), biotechnologies, and geoengineering (e.g., Currie, 2018). On the practical side, we have developed and refined our methods for expert elicitation and engagement, particularly through collaboration with the horizon-scanning and foresight strand of the programme. This also developed into work at the intersections of technology areas, such as the implications of the application of machine learning to nuclear command and control structures (Avin and Amadae, 2019). Our approaches to advancing understanding of extreme technological risk and management options in our work align well with key areas of responsible innovation frameworks: including a range of actors; seeking to anticipate future developments in science, technology, and policy, and their implications; and promoting reflection and facilitating discussion among stakeholders about key ethical issues and responsibilities arising around certain areas of research. This chapter explores different approaches to responsibility in biotechnological research and how these can complement each other, and connect to other components of management within a 'web of prevention'. Chapter 7 explores different frameworks for understanding the role activism can play in fostering responsible innovation.

Responsibilities associated with managing extreme technological risk might be understood as a subset of broader responsibilities assigned to those involved in scientific and technological research and innovation. Research and practical experience in recent years have highlighted the value of not considering such responsibilities 'in separation' or as something distinct from broader responsible conduct; instead they emphasise that embedding them in established and familiar principles, policies, and practices can assist their understanding and acceptance.

This chapter seeks to situate responsibilities for management of extreme biotechnological risk within broader approaches to responsible research and innovation, surveying both general approaches and some specific examples of initiatives concerned with the responsible management of cutting-edge life sciences research.

Two main framework concepts are used in this chapter to facilitate understanding of approaches to responsible research in the life sciences and related fields: the web of prevention, which has emerged from scholarship in international relations; and responsible research and innovation, which draws from science and technology studies. Both frameworks are useful in helping us understand how a range of approaches to supporting responsible/responsibility in life sciences research can function in complementary ways and, in combination with other areas of governance, can more effectively advance management of extreme risks. Several approaches to responsibility are outlined, and some particular cases in practice highlighted, showing the learning that can be taken from this area, and noting some of the challenges that remain.

6.2 The Web of Prevention

A useful initial concept for understanding where responsibility fits within the management of extreme technological risk is the 'web of prevention'. This has been developed within efforts to prevent the hostile application of biological and chemical sciences, and it centres on that fact that international treaty-based prohibitions on biological and chemical warfare do not stand alone and would not, by themselves, be sufficient for prevention of misuse (see, for example, Rappert and McLeish, 2007; Pearson, 1993, 2015). Instead, a web of measures is required with multiple layers and components that, functioning together, greatly improve the strength of such activities.

As well as the international prohibitions on biological and chemical weapons, and national legislation that implements those prohibitions, other components of the web of prevention include standards for the safe handling and transport of pathogens; laboratory biosafety and biosecurity; codes of conduct and ethics; biosecurity education; export controls; and preparedness and response efforts for disease outbreaks, whatever their origin. These components intersect across levels of governance from local and institutional through to global, and encompass a range of formal and informal measures. (Similarities to the concept of international regimes are explored, for example, in Rappert and McLeish, 2007, pp. 5–6.)

While developed in relation to preventing misuse, the web of prevention concept can apply to other sources of biological risks — indeed components that are designed to address areas such as natural disease outbreaks, laboratory biosafety, and environmental protection form part of the web, even when it is limited to preventing misuse (Pearson, 2015). The concept may also be useful when transferred to other areas of science and technology.

When considering approaches to responsible research, it is primarily the more informal, or 'bottom-up', parts of the web that are of interest; notably, these are also multi-component and multi-layered. A range of approaches to responsible / responsibility in science, technology, and innovation have developed over decades. These often build on each other, although some perform more specific functions or target particular audiences, actors, or parts of research and innovation systems.

6.3 Responsible (Research and) Innovation

Another useful framing concept, developed somewhat later than the web of prevention, is 'responsible research and innovation' or 'responsible innovation'. Over the last decade, an extensive body of literature and a set of practical initiatives have built up in this area. Various versions of responsible innovation have been put forward and associated frameworks have been applied, for example, by funding bodies and research institutions. Rather than drawing from all these activities, for the purposes of this chapter, the focus is on a prominent framework — outlined by Stilgoe, Owen, and Mcnaghten in 2013 — and how it highlights complementarities with other approaches to responsible science, and can add to the range of tools and methods for managing extreme biotechnological risk.

> [T]he question arises as to what are the values that underpin the scientific knowledge production system ... and to what extent these align with articulations of broader societal values and visions of the social good ... to what extent have the values and priorities tacitly embedded in scientific innovation been subjected to democratic negotiation and reflection? (Macnaghten, 2020, p. 8)

Broadly, responsible innovation (RI) developed based on awareness of and dissatisfaction with the flaws in existing models for governing science and particularly how those models understood the relationship of science to society. Critiques of particular relevance in the context of METR include a lack of openness to society being able to inform and influence innovation systems despite choices — for example in investment and regulation of innovation — having significant societal implications, as well as governance systems being mismatched to the challenges we collectively face, for example, lacking mechanisms for managing unexpected futures and for appropriately connecting with societal interests in and concerns about those futures.

In one sense, RI provides a broader approach to responsibility than those covered elsewhere in this chapter: it encompasses innovation systems as a whole and all actors and institutions that are either part of such systems or can/should exert external influence upon them. It targets policy and governance actors as much as it targets scientific and technological actors. However, RI is also intended to be embedded within and responsive to particular local and institutional contexts.

The framework outlined by Stilgoe *et al.* (2013) includes four main dimensions, seeking to move innovation systems towards models of governance that are more:

- Anticipatory — able to respond more effectively to the 'pace of social and technical change', attempting to address unforeseen 'detrimental implications of new technologies' and to achieve 'early warnings of future effects' (p. 1571).
- Reflexive — with actors and institutions involved in innovation systems able to achieve awareness of and apply some scrutiny to how their own practices, values, and assumptions are shaped by a range of factors and dynamics within and acting upon those systems, including 'being aware of the limits of knowledge and being mindful that a particular framing of an issue may not be universally held' (p. 1571).
- Inclusive — open to diverse stakeholders, including the public, to engage in the processes that shape innovation and its outcomes (particularly those which in turn will shape societal futures), with some emphasis on participatory approaches, public dialogue, and other forms of 'non-expert' engagement. This is a particularly challenging area, but ideally would enable anyone impacted by innovation systems to have some influence on their governance.
- Responsive — in response to the outcomes of anticipation, reflexivity, and inclusion, the questions and issues they raise, and to changes in external contexts, such as changing political or economic dynamics and emerging societal challenges (p. 1572). 'Responsible innovation requires a capacity to change shape or direction in response to stakeholder and public values and changing circumstances ... [and] requires attention to metagovernance — the values, norms and principles that shape or underpin policy action ...' (p. 1573).

The authors note that there will be significant interactions between the different dimensions and that attention is needed in regard to how to ensure such interactions have a positive effect. This includes a note that:

> Public dialogue, bioethics, research integrity, codes of conduct, risk management and other mechanisms may target parts of the governance of science, but they do not offer an overarching, coherent and legitimate governance approach unless we consider how they are aligned with one another. (Stilgoe *et al.*, 2013, p. 1574)

This aligns well with the web of prevention concept — we should not expect that any single approach to responsible science will achieve desired outcomes by itself; to effectively address challenges of extreme biotechnological risk, a combination of approaches, implemented in a considered way, so that they can function in a coherent and complementary manner, is needed.

6.4 Bio(techno)logical Risks of Concern

There are substantial overlaps and interactions between biotechnological risks and other biological risks and in the measures required for their effective management. Therefore, while this book and chapter are mainly concerned with technological risks, much of what is covered here is relevant to other sources of biological risks, and references in the chapter to 'biotechnological risks' can generally be understood as covering those other sources as well.

Potential sources of biological risks — affecting humans, animals, plants and ecosystems, and potentially physical infrastructure — fall within four main categories:

- Naturally occurring — within this, many threats may be, at least partially, human-induced, e.g., antimicrobial resistance or encroachment into habitats bringing increased contact with zoonotic pathogens.
- Accidents and unintentional releases.
- Deliberate releases with benign intent, but unintended or unanticipated consequences.
- Deliberate releases with malign intent.

When considering extreme risks, we are concerned about:

- Direct human casualties and fatalities from major and potentially multiple disease outbreaks.
- Indirect human casualties and fatalities from major and potentially multiple crop or livestock losses with severe impacts on food security.
- Indirect human casualties and fatalities from biological impairment or destruction of critical systems, e.g., through a novel organism that degrades components of physical infrastructure.
- Some combination of these.

A cross-cutting paper 'Classifying Global Catastrophic Risks' developed early in the METR programme provides a complementary and more extensive exploration of the contributions of biological components to this level of risk (as 'replicators' and 'critical systems'). It also notes the fragilities in prevention and mitigation that can arise from, for example, institutional cultures, and individual and interpersonal behaviours (Avin, *et al.*, 2018, Fig. 3, p. 23), to which the responsibility approaches outlined in this chapter are directed.

6.5 Interaction of Scientific and Technological Advances and Biotechnological Risks

Historic biological warfare programmes have rapidly incorporated scientific and technological advances. Such advances can alter the profile of misuse, for example, by extending the range of physiological targets for hostile use or reducing the skill level and sophistication of facilities and equipment needed. For these reasons, attention is paid to improving understanding about the ways in which such advances can both facilitate and mitigate against misuse (see, for example, IAP, 2016a; Warmbrod *et al.*, 2020). Studies in this area may focus on particular instances of technological development, such as CRISPR Cas-9 as a gene editing tool (West and Gronvall, 2020; Kirkpatrick *et al.*, 2018); but it is possibly the trends in technology — such as gene editing generally becoming more accurate, cheaper, and more widely used — that are more timely guides to management needs. Trends in technological convergence are also important, for example, developments in nanotechnology may provide novel delivery mechanisms for biological agents.

Generally, scientific and technological (S&T) advances are making deliberate harmful applications and benign applications with severe unintended consequences easier to achieve. Also, as knowledge and data — such as from genome sequencing — are globally disseminated, capacities for harm are also more widespread. This is highlighted, for example, on p. 2 of the Joint NGO Statement to the Biological Weapons Convention Meeting of Experts in 2021:

> [D]ecreasing cost and widespread access to advanced capabilities continue to lower barriers to utilizing many of these technologies and associated pathogens … the rapid development of S&T may be outpacing necessary ethical and regulatory practices, and governments and civil society often struggle to anticipate emerging capabilities and to identify and implement appropriate oversight mechanisms.

Such interactions between S&T advances and extreme risks occur across many emerging fields and the need for greater understanding of likely future scenarios in a way that can usefully guide current management activities motivates the horizon-scanning and foresight strand of CSER's ongoing work (see Chapter 4), which has included two exercises in biological engineering (Wintle *et al.*, 2017; Kemp *et al.*, 2020).

6.6 Approaches to Responsible/Responsibility in Life Sciences Research

6.6.1 *Responsible conduct of research*

Responsible conduct of research (RCR) developed in the US in the 1980s in response to a perceived need to formalise training in ethical research conduct for scientists, particularly in the biomedical field (Steneck and Bulger, 2007). The foundations and core principles of RCR are covered in the US National Academies publication *On Being A Scientist: A Guide to Responsible Conduct in Research*, first published in 1989. RCR is utilised by a range of professional associations and research institutions, often being an expected part of early career training. Its scope has extended over time and it has more recently been presented in the global context in *Doing Global Science: A Guide to Responsible Conduct in the Global Research Enterprise* (IAP, 2016b), which aims to promote internationally harmonised standards in RCR. RCR includes areas such as appropriate treatment of data and reporting of results; addressing misconduct; treatment of human participants and animal subjects; safety; authorship and publication; intellectual property rights; and handling of conflicts of interest (NRC, 2009a).

RCR training is intended to complement informal transfer of values, principles, behaviours, and good practices attained through work with more senior scientists, with science academies recognising that — particularly due to changes in scientific research structures and the higher profile of science–society interactions — such informal dissemination is likely to be insufficient by itself (NRC, 1989, preface).

The RCR approach focuses primarily on what might be termed the internal practice of science, concentrating on how science is practiced within educational and professional contexts, in order to uphold its integrity. It also addresses external considerations of the implications for society, the social role of science, and the relationship between science and the public, recognising the complexity of 'the problems facing modern society' where science needs to interact with social and political processes:

> The standards of science extend beyond responsibilities that are internal to the scientific community. Researchers also have a responsibility to reflect on how their work and the knowledge they are generating might be used in the broader society. (NRC, 2009a, p. 48)

6.6.1.1 *Responsible conduct of research and extreme technological risk*

Having scientific communities and individual scientists attuned to ethical responsibilities related to their work and recognising the key role that value judgements play in science provide an important basis for engagement with and understanding of more specific responsibilities associated with research that has the potential to pose extreme risks. Consideration of the external, societal aspects of RCR is important in terms of establishing an expectation of scientists to be aware of and contribute to responses to extreme technological risk: these are an urgent and compelling societal concern.

In more recent editions of *On Being A Scientist* and in related work by the National Academies and partner organisations, there has been increasing attention to the pedagogical approach taken to RCR education, and this has been applied within many of the biosecurity and ethics education initiatives covered in the next section.

RCR forms a key component of scientific training by providing awareness, alertness, and some understanding of the needs and a core motivation for managing risks associated with research, including extreme risks. It can be usefully supplemented by approaches that address some of the risks more specifically and by those that extend consideration to other parts of research and innovation systems. Placed within a global context that continues to have substantial disparities in scientific and technological capacities and in the distribution of benefits and risks from technology, it is also essential to incorporate consideration of these dynamics and how to respond to them.

The RCR approach provides a clear connection to the responsibilities of individuals, complementing some of the other approaches where the responsibilities are more easily assigned to communities or institutions and do not always readily translate to individual actions. RCR also makes clear that there are instances of misconduct that damage individuals, institutions, and public trust, and can harm society, and that all scientists have responsibilities to address these. This would apply to any deliberate misuse of biotechnology, and also to instances in which, for example, safety risks are not being appropriately addressed. The ability to identify such cases as misconduct requires awareness of how and where such risks arise, something that is addressed by some of the more specific approaches to responsibility outlined in the following.

On Being A Scientist and many other presentations of RCR in professional training generally do not specifically consider catastrophic risks and deliberate misuse; however, they clearly point to the justification for paying attention to these areas as part of scientific responsibilities, and they provide an important foundation for incorporating these more specific topics into scientific education, training, and professional standards. This is clear, for example, in the contributions of science academies to the development and implementation of training in areas such as dual-use research and biosecurity.

The 2016 publication *Doing Global Science*, which outlines a global approach to RCR, does explicitly address misuse and the responsibilities of scientists to address technological risk. It sets out expectations that researchers will participate in discussions and forums exploring issues relating to 'preventing the deliberate misuse of research' and 'the possible consequences of their work, including harmful consequences' (IAP, 2016b, p. 22); and that this needs to be done from the early stages of training and within the planning for research projects (pp. 22–23). It further outlines responsibilities of researchers to participate 'in the creation of institutions and practices to address the possible risks of existing and emerging technologies' (p. 22) and 'to contribute to the development and dissemination of these standards' (p. 5).

6.6.2 *Education, awareness-raising, and codes of conduct for prevention of misuse of the life sciences*

For scientists to be able to fulfil responsibilities related to preventing misuse of research, they need to have a good level of awareness and understanding of what those risks are, why they are of concern, and of the ways in which they can contribute to managing them. (The same is true of risks relating to accidents — this is covered in the next section.) Several studies in the 2000s demonstrated limited awareness among scientific communities of the Biological Weapons Convention, the risks of misuse, and their responsibilities in this regard (Gaudioso and Zemlo, 2007; Kelle, 2007; Mancini and Revill, 2008; Minehata and Shinomiya, 2009; Minehata, 2010; and NRC, 2009b). This prompted a range of activities to improve the situation, including the development of educational materials, training, and codes of conduct. These activities have been supported by science academies, academic and civil society groups, and by states parties to the Biological Weapons Convention. Work on education and training has incorporated research into effective pedagogical approaches and advice on implementing these, for example, through team-based and other active learning approaches. There is also wide recognition of the need for educational activities and materials to be adapted to local contexts.

6.6.2.1 *Codes of conduct*

Within the context of the Biological Weapons Convention (BWC), the development, adoption, and dissemination of codes of conduct have been promoted as valuable tools for raising awareness of responsibilities for the prevention of misuse among research communities. BWC review conferences and meetings have actively encouraged the use of such codes as part of national implementation measures. While they may be usefully developed and deployed by any research institution and adapted to national and local contexts, it has also been considered important to undertake collaborative work to provide a consensus framework on an international basis, with extensive input from the scientific community.

From this work, the *Tianjin Biosecurity Guidelines for Codes of Conduct for Scientists* have recently been published. These were developed through a collaboration involving the Interacademy Partnership, Tianjin University's Center for Biosafety Research and Strategy, and the John Hopkins Center for Health Security; it is hoped that these guidelines will soon be endorsed by states party to the BWC and that their dissemination will form part of their national education and outreach activities (*Joint NGO Statement*, 2021, p. 3). The *Guidelines* clearly build on the RCR concept — explicitly including it and connecting it directly to preventing misuse of research (Tianjin University, JHCHS — 2021, *Tianjin Guidelines*, p. 1). Similarly to RCR, the guidelines address individual, institutional, and collective responsibilities and the need to sustain public trust through engagement, for example, in statements 8 and 9 on p. 2:

> Scientists should advocate for peaceful and ethical applications of the biosciences and work collectively to prevent misuse of biological knowledge, tools and technologies ...

> [Institutions] should be aware of the potential for misuse ... and ensure that expertise, equipment and facilities are not used for illegal, harmful or malicious purposes ... should establish appropriate mechanisms and processes to monitor, assess, and mitigate potential vulnerabilities and risks ... and establish a training system for scientists.

The responsibility of scientists to 'identify and manage potential risks when they pursue the benefits of biological research and processes' and 'consider potential biosecurity concerns' extends to all stages of research (statement 5, p. 2).

In addition to the links to RCR, the *Guidelines* refer to rules and mechanisms that should be in place to 'prevent, mitigate and respond to risks' and recommend establishment of 'a culture of safety and security' (*ibid.*). These areas connect to work of the World Health Organization (WHO) and World Organisation for Animal Health (WOAH) on laboratory biosafety and biosecurity.

6.6.2.2 *Biosecurity education and training*

Biosecurity education and training also has an awareness-raising function similar to codes of conduct but aims to extend beyond that to an understanding of the risks of misuse and the development of competencies that would, for example, enable identification of areas of particular risk and incorporation of ethical reflection into decision-making in research. The FutureLearn course 'Next Generation Biosecurity: Responding to 21st Century Biorisks', for example, aims to enable students to:

- Understand and explain the need for laboratory biosafety and biosecurity, and what these entail.
- Discuss how biosecurity policies, practices and guidelines are developed at international and national levels and how to contribute to these processes.

- Summarise what it means to be safe in the lab, how to apply these principles and what to expect from others to keep you and the community safe from accidental and deliberate actions.
- Describe the biosecurity challenges facing the 21st century, especially as the revolution in life sciences continues.
- Identify areas of security and ethical concern within the life sciences and reflect on strategies and approaches to deal with them as they arise.

The development of such education programmes has also been encouraged by review conferences and meetings of the BWC. The value of such work was also highlighted within the 2018 *UK Biological Security Strategy* (HM Government, 2018, p. 22).

Several groups have led efforts to develop and deliver educational materials such as textbooks, online courses, and train-the-trainer modules (examples include: Whitby *et al.*, 2015; Millett and Edwards, undated; Gryphon Scientific, undated; and University of Bradford, undated). These have often involved collaboration between academic groups and science academies, and efforts have been made to understand and address potential barriers to their implementation in different contexts, and to promote work for local adaptation (for example, see NRC, 2013; Muneer *et al.*, 2021). The importance of adaptation to local contexts is part of the responsible innovation (RI) framework and is also recognised throughout the WHO's work on biorisk management.

The content of courses varies, but generally includes reference to the prohibition on misuse within the BWC, an overview of past misuse, RCR and related ethical principles, and cases which help explore issues of potential misuse of research. It is recognised as important that such training not only conveys information but also equips learners with skills and competencies that will allow them to apply their knowledge as they come across such issues in their future work. This aligns with the growing understanding of the value of active learning approaches (Knight and Wood, 2005; NRC, 2011), which has been applied in the development and delivery of biosecurity education.

Active learning approaches are generally student-centred and often involve a combination of collaborative (group) learning and self-directed study. The incorporation of such approaches helps to achieve the reflexive practices promoted by responsible innovation frameworks. Novossiolova (2016), for example, demonstrates how team-based learning approaches can support biosecurity education. She notes, based on experience of delivery in several countries, that team-based learning provides:

> an easy-to-replicate, user-friendly approach, that can be applied in many different education settings at various stages of instruction … [and] enables the instructor to cover new material in a way that engages learners as active participants, allowing them to take ownership of their own learning, and develop reflection and self-evaluation skills. (p. 2)

There are similar educational activities developed in relation to the Chemical Weapons Convention and there has been learning across these initiatives.

6.6.2.3 *Contributions to addressing extreme biotechnological risk*

These initiatives form a key part of the web of prevention, enabling scientists to understand risks of deliberate misuse of research, and consider and discuss these within their institutions and communities, and with publics and policymakers. Without this understanding, responsibilities are unlikely to be fulfilled. Of course, it is recognised that education does not guarantee adoption of appropriate behaviours, but nor can responsible practice be expected in the absence of a good understanding of what those responsibilities are and why they are important. In addition, to be able to apply understanding of these responsibilities in future work, it is important that skills and competencies are developed in previously unfamiliar areas such as ethical decision-making around risks and benefits of research.

The most extreme biological risks are now likely to come from hostile application of advances in biotechnology and related fields with the potential to cause greater devastation than naturally occurring disease outbreaks. As capabilities to apply biotechnology in such ways become increasingly widespread, education and awareness-raising activities need to be similarly widespread so that scientists, individually and collectively, are able to manage risks of research being misused in such potentially devastating ways. This needs to happen from the earliest stages of training, because later work may take place in non-traditional settings in which formal oversight mechanisms are lacking.

6.6.3 *Work of the World Health Organization (WHO) and World Organisation for Animal Health (WOAH)*

> Safe and secure working practices associated with the conduct of research in laboratory settings are important elements for addressing the risks that could potentially arise from accidents or the deliberate misuse of life sciences research. Good laboratory biosafety practices will mitigate the risks posed by laboratory accidents while laboratory biosecurity procedures will strengthen the accountability and responsibility of laboratory workers and their managers and thereby enhance public confidence in the responsible conduct of scientific experiments. (WHO, 2010, p. 31)

As mentioned earlier in this chapter, there are many commonalities in efforts to address technological and non-technological sources of biological risks. The BWC and its associated meetings act as the main international forum for upholding the prohibition against use of biological sciences for hostile purposes, but other international organisations also have an interest in ensuring prevention of misuse and responsible practice in and utilisation of life sciences research, particularly the three organisations with mandates to protect human, animal, and plant life and health: the WHO, WOAH, and Food and Agriculture Organization (FAO).

The WHO and WOAH primarily focus on control of naturally occurring disease outbreaks, but they also oversee a range of activities relevant to minimising accidental releases of pathogens from laboratories and during transport, and have more recently extended this work to biosecurity and the inclusion of concerns about dual-use research in responsible scientific practice. The WHO has brought this work together in its biorisk management framework, and WOAH through its *Biological Threat Reduction Strategy* and associated activities (https://www.woah.org/en/what-we-do/global-initiatives/biological-threat-reduction/).

Work on biosafety brings technical standards and practical guidelines to the safety dimensions of RCR. As with other areas of RCR, safe practices serve both internal and external purposes — they safeguard individual researchers and their colleagues, protect the wider public and environment, and help to maintain public trust. The WHO's main publications on biosafety are its *Laboratory Biosafety Manual* and its *Guidance on Regulations for the Safe Transport of Infectious Substances*. The aim is to prevent accidents that can harm those working in laboratories and in the handling and transport of pathogens, and accidental releases into the environment. (Once an accidental release occurs, there are other mechanisms of the WHO that can address subsequent stages of surveillance and response, should this become necessary.)

The WOAH also provides standards relating to safety in veterinary laboratories and in the transfer and handling of pathogens within its *Terrestrial Animal Health Code, Aquatic Animal Health Code*, and associated Manuals. Guidance on biosafety and biosecurity issues can, for example, be found in: Chapter 5.8 of the Terrestrial Code — International Transfer and Laboratory Containment of Animal Pathogenic Agents (WOAH, 2021a); and in Chapter 1.1.4 — Biosafety and Biosecurity: Standard for Managing Biological Risk in the Veterinary Laboratory — in the Terrestrial Manual (WOAH, 2021b). The latter is accompanied by some worked examples in Chapter 2.1.3. Similar to the WHO, the WOAH also has standards, processes, and mechanisms in place that would assist surveillance and response should an accidental release occur.

In the 2004 edition of the *Laboratory Biosafety Manual* (WHO, 2004), the WHO added a chapter on laboratory biosecurity, and two years later it published a more extensive document *Biorisk Management: Laboratory Biosecurity Guidance* (WHO, 2006). Laboratory biosecurity is differentiated from laboratory biosafety as follows (WHO, 2006, pp. iii–iv):

> Laboratory biosafety describes the containment principles, technologies and practices that are implemented to prevent the unintentional exposure to pathogens and toxins, or their accidental release.

> Laboratory biosecurity describes the protection, control and accountability for valuable biological materials … within laboratories, in order to prevent their unauthorized access, loss, theft, misuse, diversion or intentional release.

The *Guidance* also introduced the biorisk management concept, '[t]he analysis of ways and development of strategies to minimize the likelihood of the occurrence of biorisks' (WHO, 2006, p. iii), which combines laboratory biosafety, laboratory biosecurity, and bioethics. It aims to:

> develop a comprehensive laboratory biosafety and biosecurity culture, allowing biosafety and biosecurity to become part of the daily routine

of a laboratory, improving the overall level of working conditions, and pushing for expected good laboratory management. (WHO, 2006, p. 11)

The approach is specifically designed to be adaptable to local situations, outlining 'recommendations and performance expectations' that 'do not call for compliance with a set of requirements, but rather to help identify and set goals to be achieved', enabling 'countries and facility managers to define and choose appropriate systems and controls to ensure that the biorisk management goals that have been identified are reached. It allows institutions to adapt their local biosecurity plans to their particular situation' (*ibid.*). This aligns with the approach taken by education initiatives mentioned in the previous section, which also recognise the importance of adaptability to local contexts, and aligns with the RI framework.

When applied to pathogens and toxins, the *Guidance* outlines expectations that 'laboratories holding them should address the dual-use nature of such agents and take responsibility … for deciding on the adoption of appropriate biorisk management measures to protect them' (WHO, 2006, p. 16), and — in a specific section addressing potential misuse of bioscience (5.2) — that 'the potential for harmful misuse may suggest the need for specific protective measures for laboratory facilities, the VBM they contain, the work performed and the staff involved'.

Fitting well with the approaches to education and awareness-raising, among the main components of biorisk management programmes are promotion of 'a culture of awareness, shared sense of responsibility, ethics and respect of codes of conduct within the international life science community' (WHO, 2006, p. 30).

While the fourth edition of the *Laboratory Biosafety Manual* does not refer to the biorisk management approach, it refers to the need for countries, institutions, and laboratories to consider 'ethical issues related to life sciences in a risk assessment', and specifically in the context of research 'with emerging technologies, for which limited information currently exists, the scientific community must: promote a culture of integrity and excellence, distinguished by openness, honesty, accountability and responsibility' (WHO, 2020, p. 88). In line with the anticipatory component of responsible innovation, there is also an expectation of seeking to understand the potential risks and impacts of emerging technologies, updating oversight and risk assessment 'as additional information is obtained over time' and to 'monitor and assess the scientific, ethical and social implications of certain biotechnologies, and, as warranted, monitor the development of those technologies and their integration into scientific and clinical practice' (WHO, 2020, p. 89).

The biorisk management approach forms a core part of the WHO's 2010 publication *Responsible Life Sciences Research for Global Health Security*. The biorisk management framework in this publication incorporates three pillars:

- Research excellence — 'research that is of high quality, ethical, rigorous, original and innovative' (p. viii);
- Ethics — 'the promotion of responsible and good research practices, the provision of tools and practices to scientists and institutions that allow them to discuss, analyse and resolve in an open atmosphere the potential dilemmas they may face in their research' (p. 2);
- Biosafety and laboratory biosecurity.

The ethics pillar clearly fits well with the reflexive component of RI, and the overall approach again aligns with the RI recommendation of local adaptation: 'This guidance provides Member States with a conceptual framework for individual adaptation according to national circumstances, contexts, needs and capacities'. It connects this to variability in the availability of resources in different settings as well as different local needs, one of these being not to overburden regulatory capabilities (p. 20). Biorisk management policies should be:

> flexible to incorporate new scientific developments; sustainable in order to meet the differing needs of countries and institutions; viable for countries facing competing demands with scarce resources; developed in collaboration with relevant stakeholders, particularly researchers who are the most directly affected by the policy, so that it is acceptable and equitable to all stakeholders; and built on existing frameworks and experiences. (p. 20)

This also picks up on the RI components of anticipation, inclusion, and responsiveness.

The publication draws in elements from RCR, for example, stating that 'students, researchers and laboratory staff need to receive appropriate education and training on ethics and best practices in responsible conduct of research' (p. 20), and outlines the need for scientists to understand and engage with discussions around dual-use issues and take action to prevent misuse of research (and generally maintain high ethical standards in research).

Linking consideration of ethics to issues of biosafety and biosecurity — which had often been separated previously, for example, in terms of institutional review processes — is viewed to be important given: 'the potential conflicting values of promoting scientific progress and protecting public security, and the questions about responsibility that arise'; 'the potential dangers to the environment and society'; and, overall, that the challenge of managing potential risks while realising the beneficial potential of life sciences research is not just a technical, but also an ethical challenge (WHO, 2010, p. 25).

The 2010 publication frames its guidance primarily for biomedical researchers, but the WHO is currently convening diverse expertise to develop a guiding framework on responsible life sciences research, which will have more of a member state focus — see WHO (2021).

In 2015, the WOAH published its *Biological Threat Reduction Strategy* recognising that its work to protect and promote animal health necessarily includes addressing threats of accidental release or deliberate misuse (WOAH, 2015, p. 2). In describing such threats, the *Strategy* explains:

> These "unnatural" biological threats carry special risks because pathogens may be engineered or released in such a way as to make them more harmful. Although the probability of a deliberate or accidental release may be relatively low, the impact may be catastrophic from a national to a global level.

> Animal pathogens may be used as bioweapons, in bio crime or in bioterror because they have a high impact, are cheap, easy to acquire and propagate, and can be readily smuggled through border checks undetected. The biotechnology revolution means that options for engineering animal pathogens are increasing all the time (and becoming more widely available), whilst the cost of doing so is decreasing. (p. 2)

The *Strategy* recognises the need to integrate expertise in biosafety, biosecurity, bioethics, and biotechnology; to create a culture of responsible science; for biothreat awareness (referring to accidental release and deliberate misuse risks); for capacity building in various areas, including for policymakers as well as veterinarians; and to incorporate biothreat reduction into veterinary education programmes (pp. 6–7). The *Strategy* also understands the WOAH's responsibilities to include coordination with other international bodies and initiatives and to 'advocate for the non-proliferation of biological weapons including support for the [Biological Weapons Convention]' (p. 7).

The *Strategy* has been complemented by additional guidance materials produced by the WOAH on responsible conduct in veterinary research; investigation of suspicious biological events; and biosafety and biosecurity in veterinary laboratories and animal facilities. Its 2019 *Guidelines for Responsible Conduct in Veterinary Research: Identifying, Assessing and Managing Dual Use*, for example, aim to:

> raise awareness about the dual-use potential of research in veterinary settings, supporting veterinary professions, researchers and other stakeholders to effectively identify, assess and manage dual-use implications. (p. 2)

The *Guidelines* also provide a list (Box 3, p. 5) to 'trigger critical thinking about intended and potential unintended consequences of research' including:

- Could the research alter the distribution of an animal or plant species in the environment with harmful effect?
- Could the research result in a new or recreated pathogen or toxin?
- Could the research result in reduced host immunity, increased host susceptibility, and/or altered host tropism?
- Could the research promote or induce resistance to therapeutic or prophylactic measures?
- Could the research interfere with detection or diagnosis of a microorganism or toxin?
- Could the research alter the nature of an animal feed or feed plant with harmful effect?
- Could the context or manner in which the research is published or otherwise communicated facilitate misuse? (p. 5)

This is another example of the promotion of reflexivity that aligns well with the RI framework. The *Guidelines* also connect with RCR and the WHO's biorisk management approach: 'Responsible conduct includes considerations related to safety, security and ethics. Recognising the dual-use implications of research is an integral component of being a responsible scientist' (WOAH, 2019, p. 5).

Section three of the *Guidelines* explicitly connects handling of dual-use issues to RCR, and outlines examples of the related responsibilities of different stakeholders through the research process. Fitting with the inclusive component of RI, this includes a statement that:

> Institutions should engage with the wider community, as both the potential benefits and risks of research ultimately affect the entire society. Mechanisms to include the general public in the consultation process are recommended for selected projects or activities. (WOAH, 2019, p. 5)

And, fitting with the anticipatory component, it includes a statement that:

> Regulatory authorities often struggle with the fact that scientific advances outpace regulatory frameworks for dealing with research. This can create new safety, security or ethical issues with respect to animal and human health and the environment. Horizon scanning should be an integral part of regulatory frameworks in order to help anticipate new developments and react in a timely manner. (WOAH, 2019, p. 6)

6.6.4 *How these assist management of extreme biotechnological risk*

The involvement of the WHO and WOAH in efforts to promote responsible scientific practice connected to prevention of misuse, as well as other biological risks, is important because it expands the actors who become aware of these issues and of their relevance to their work. It makes clear that these are not just concerns relevant to the security community to be handled solely within forums on arms control and non-proliferation. This is significant because those who work on human, animal, and zoonotic diseases in both research and clinical settings (and are, for example, members of the WHO and WOAH research networks) are particularly important to reach and engage in efforts that support responsible research practise — novel research on human and animal disease can carry particularly high risks associated with accidental release and deliberate misuse, alongside potentially transformative benefits. In a similar way to building biosecurity education from RCR, the joining of biosecurity concerns to more familiar principles and practices of biosafety can assist understanding and acceptance. For both areas, making the link with bioethics can provide a route into discussion of difficult issues around balancing risks and benefits.

6.7 Conclusion and Challenges Remaining

For extreme risks associated with existing and emerging technologies, the actions and attitudes of scientists towards the risks associated with their work have the potential to drive and heighten those risks or to contribute to safeguarding humanity. Initiatives to promote responsible science therefore constitute a key part of overall activities to manage extreme technological risk. In the area of biotechnological risks, that overall landscape has been described as a 'web of prevention' in which we hope to work towards positive interactions between different components, which in combination will guide us towards beneficial outcomes for humanity. Several specific initiatives — including activities by science academies, civil society groups, academics from a range of disciplines, and international organisations — have been developed to promote responsible science within biotechnology, the life sciences, and related fields. Each of these has value and there is substantial complementarity among them and the contributions they can make to managing extreme technological risk.

Unsurprisingly, there are many challenges faced in effectively achieving these contributions. Some relate to the need for further investment and resourcing to promulgate codes, standards, and other education and awareness-raising activities; enable their adaptation to local contexts; and broaden the engagement of all stakeholders in their development and implementation. Concern about the capacity for achieving space in crowded academic curricula to adequately cover these topics and facilitate skills development has been raised within projects to develop educational modules and materials.

The extreme nature of the risks themselves and the degree of uncertainty and expert disagreement around it also present challenges. For example, it can be challenging for researchers to recognise and act on relatively low level risks associated with their work; achieving understanding and acceptance of an association of their work with extreme risk is even more demanding. This is particularly the case in a context in which there have been problematic understandings of risk (e.g. in the regulation of genetic engineering), which have been perceived as obstructing lines of research. It is also frequently a context in which the benefits of the work may be much clearer, immediately present and understandable to those involved, and where an assumption of benign intent is deeply embedded in community culture:

> If researchers have undertaken their work for the benefit of humankind, what is the relevance for them of security or the national and international measures to promote it? Rejection or resentment is a plausible response to the suggestion that their research could pose security risks, particularly if it means additional administrative costs or requirements for a problem they may not recognize or accept as a legitimate concern. (Husbands, 2018)

Given how challenging some aspects of responsible science applied to extreme risks and prevention of misuse of research can be, a key element of the approaches outlined in this chapter is that none of them present this as a challenge to be faced by individual scientists alone — they embed these responsibilities within institutional contexts; engage the support of scientific societies and interdisciplinary communities of expertise; provide forums for learning, discussion, and critical reflection; connect to policy environments; and recognise the responsibility of other actors, for example, in forging responsible governance and achieving meaningful engagement of societal stakeholders in moving towards desirable futures.

The strand of the METR research programme focusing on responsibility has benefited from interactions with other parts of the programme, for example, the work done in horizon scanning has strong connections to the anticipatory dimension of responsible innovation, and is a key way of bringing together diverse actors to engage with efforts to manage extreme risks.

References

Avin, S. and Amadae, S. (May 2019). Autonomy and machine learning at the interface of nuclear weapons, computers and people. In V. Boulanin (ed.) *The Impact of Artificial Intelligence on Strategic Stability and Nuclear Risk*, Vol. I, Chapter 13. Stockholm: SIPRI. Available at: https://www.sipri.org/sites/default/files/2019-05/sipri1905-ai-strategic-stability-nuclear-risk.pdf (Accessed 4 December 2021).

Avin, S., Wintle, B. C., Weitzdoerfer, J., OhEigeartaigh, S. S., Sutherland, W. J., and Rees, M. J. (September 2018). Classifying global catastrophic risks. *Futures*, 102, 20–26. Available at: https://doi.org/10.1016/j.futures.2018. 02.001 (Accessed 4 December 2021).

Currie, A. (September 2018). Geoengineering tensions. *Futures*, 102, 78–88. Available at: https://doi.org/10.1016/j.futures.2018.02.002 (Accessed 4 December 2021).

Gaudioso, J. and Zemlo, T. (2007) Survey of bioscience research practices in Asia. Implications for biosafety and biosecurity. *Appl. Biosaf.*, 1294. Available at: https://doi.org/10.1177/153567600701200408 (Accessed 4 December 2021).

Gryphon Scientific Training materials for practical implementation of laboratory biosafety, biosecurity, and biorisk management. Available at: https://www. gryphonscientific.com/resources/training-materials-for-practical-implementa- tion-of-laboratory-biosafety-biosecurity-and-biorisk-management/ (Accessed 4 December 2021).

HM Government. (2018). UK biological security strategy. Available at: https://www.gov.uk/government/publications/biological-security-strategy (Accessed 4 December 2021).

Husbands, J. L. (2018). The challenge of framing for efforts to mitigate the risks of 'dual use' research in the life sciences. *Futures*, 102, 104–113. Available at: https://doi.org/10.1016/j.futures.2018.03.007 (Accessed 4 December 2021).

IAP. (2016a). Technical report — The biological and toxin weapons convention: Implications of advances in science and technology. Available at: https:// royalsociety.org/-/media/policy/projects/biological-toxin-weapons-convention/ biological-weapons-technical-document.pdf (Accessed 4 December 2021).

IAP. (2016b). Doing global science: A guide to responsible conduct in the global research enterprise. Available at: https://www.interacademies.org/ publication/doing-global-science-guide-responsible-conduct-global-research- enterprise (Accessed 4 December 2021).

Joint NGO Statement to Biological Weapons Convention Meeting of Experts, Geneva, 30 August–8 September, 2021. Available at: https://documents. unoda.org/wp-content/uploads/2021/09/Joint-NGO-Statement-v.7.pdf (Accessed 4 December 2021).

Kelle, A. (November 2007). Synthetic biology and biosecurity awareness in Europe, Bradford science and technology report, No. 9. Available at: https:// bradscholars.brad.ac.uk/handle/10454/804 (Accessed 4 December 2021).

Kemp, L., *et al.* (2020). Point of view: Bioengineering horizon scan 2020. *eLife*, 9, e54489 DOI: 10.7554/eLife.54489.

Kirpatrick, J., Koblentz, G.D., Palmer, M.J., Perello, E., Relman, D.A., and Denton, S.W. (December 2018). Editing Biosecurity: Needs and strategies for governing genome editing. Available at: https://mars.gmu.edu/server/api/core/bitstreams/4f160299-2317-4039-9e3f-d454b1092f20/content (Accessed 29 January 2024).

Knight, J. K. and Wood, W. B. (2005). Teaching more by lecturing less. *Cell Biol. Educ.*, 4(4), 298–310.

Macnaghten, P. (2020). *The Making of Responsible Innovation* (Elements in Earth System Governance). Cambridge: Cambridge University Press. Available at: https://doi.org/10.1017/9781108871044 (Accessed 4 December 2021).

Mancini, G. and Revill, J. (2008). Fostering the biosecurity norm: Biosecurity education for the next generation of life scientists. University of Sussex. Available at: http://sro.sussex.ac.uk/id/eprint/39517/1/Fostering.pdf (Accessed 4 December 2021).

Millett, K. and Edwards, B. Next generation biosecurity: Responding to 21st century biorisks. FutureLearn. Available at: https://www.futurelearn.com/courses/biosecurity (Accessed 4 December 2021).

Minehata, M. (2010). An investigation of biosecurity education for life scientists in the Asia-Pacific region Available at: https://internationalbiosafety.org/wp-content/uploads/2019/08/Biosecurity-Education-Asia-Pacific.pdf. (Accessed 4 December 2021).

Minehata, M. and Shinomiya, N. (2009). Dual-use education in life-sciences degree courses in Japan: Survey report. Available at: https://www.brad.ac.uk/acad/sbtwc/dube/publications/JapanSurvey.pdf (Accessed 4 December 2021).

Muneer, S., Afzal Kayani, H., Ali, K., Asif, E., Rehmat Zohra, R. and Kabir, F. (2021). Laboratory biosafety and biosecurity related education in Pakistan: Engaging students through the Socratic method of learning. *J. Biosaf. Biosecur.*, 3(1), 22–27. https://doi.org/10.1016/j.jobb.2021.03.003.

Novossiolova, T. (2016). *Biological Security Education Handbook: The Power of Team-Based Learning*, ISBN Number: 978 1 85143 278 3. Available at: https://www.bradford.ac.uk/media-v8/site/news/archive/Biological-Security-Education-Handbook-The-Power-of-Team-Based-Learning-(PDF,-429kb).pdf (Accessed 4 December 2021).

NRC. (1989). On being a scientist: A guide to responsible conduct in research, first edition. Available at: https://www.pnas.org/content/pnas/86/23/9053.full.pdf (Accessed 4 December 2021).

NRC. (2009a). *On Being A Scientist: A Guide to Responsible Conduct in Research* (3rd edn.). Available at: https://www.nap.edu/catalog/12192/on-being-a-scientist-a-guide-to-responsible-conduct-in (Accessed 4 December 2021).

NRC. (2009b). A survey of attitudes and actions on dual use research in the life sciences. Available at: https://www.nap.edu/catalog/12460/a-survey-of-attitudes-and-actions-on-dual-use-research-in-the-life-sciences (Accessed 4 December 2021).

NRC. (2011). Promising practices in undergraduate science, technology, engineering and mathematics education: Summary of two workshops. Available at: https://www.nap.edu/read/13099/chapter/1 (Accessed 4 December 2021).

NRC. (2013). Developing capacities for teaching responsible science in the MENA region: Refashioning scientific dialogue. Available at: https://www.nap.edu/read/18356/chapter/1 (Accessed 4 December 2021).

Pearson, G. S. (1993). Prospects for chemical and biological arms control: A web of deterrence. *Wash. Q.*, 16(2), 145–162. Available at: https://doi.org/10.1080/01636609309443401 (Accessed 4 December 2021).

Pearson, G. S. (2015). The idea of a web of prevention, Chapter 7. In S. Whitby, *et al.* (eds.), *Preventing Biological Threats: What You Can Do — A Guide to Biological Security Issues and How to Address Them*. Available at: https://bradscholars.brad.ac.uk/handle/10454/7821 (Accessed 4 December 2021).

Rappert, B. and McLeish, C. (eds.). (2007). *A Web of Prevention: Biological Weapons, Life Sciences and the Governance of Research* (1st edn.). London: Earthscan.

Steneck, N. H. and Bulger, R. E. (September 2007). The history, purpose, and future of instruction in the responsible conduct of research. *Acad. Med.*, 82(9), 829–34. doi: 10.1097/ACM.0b013e31812f7d4d.

Stilgoe, J., Owen, R., and Macnaghten, P. (2013). Developing a framework for responsible innovation. *Res. Policy*, 42, 1568–1580.

Tianjin University, John Hopkins Center for Health Security, and the Interacademy Partnership. (2021). Tianjin biosecurity guidelines for codes of conduct for scientists. Available at: https://www.centerforhealthsecurity.org/our-work/Center-projects/IAPendorsementTianjinCodes/20210707-IAP-Tianjin Guidelines.pdf (Accessed 4 December 2021).

University of Bradford. Applied dual-use biosecurity education module. Available at: https://www.bradford.ac.uk/bioethics/educational-module-resources-emr/english-language-version-of-educational-module-resource-emr/ (Accessed 29 January 2024)

Warmbrod, K. L., Revill, J., and Connell, N. (2020). *Advances in Science and Technology in the Life Sciences: Implications for Biosecurity and Arms Control*. Geneva: UNIDIR. Available at: https://unidir.org/publication/advances-in-science-and-technology-in-the-life-sciences/ (Accessed 29 January 2024).

West, R. M. and Gronvall, G. K. (Winter 2020). CRISPR cautions: Biosecurity implications of gene-editing. *Perspect. Biol. Med.*, 63(1), 73–92. doi:10.1353/pbm.2020.0006.

Whitby, S., Novossiolova, T., Walther, G., and Dando, M. (eds.). (2015). *Preventing Biological Threats: What You Can Do — A Guide to Biological Security Issues and How to Address Them*. Available at: https://bradscholars. brad.ac.uk/handle/10454/7821 (Accessed 4 December 2021).

WHO. (2004). *Laboratory Biosafety Manual* (3rd edn.). Available at: https://www. who.int/publications/i/item/9241546506 (Accessed 4 December 2021).

WHO. (2006). *Biorisk Management: Laboratory Biosecurity Guidance*. Available at: https://www.who.int/publications/i/item/biorisk-management-laboratory-biosecurity-guidance (Accessed 4 December 2021).

WHO. (2010). *Responsible Life Sciences Research for Global Health Security*. Available at: https://apps.who.int/iris/handle/10665/70507?search-result=true &query=responsible+life+sciences+research&scope=&rpp=10&sort_ by=score&order=desc (Accessed 4 December 2021).

WHO. (2020). *Laboratory Biosafety Manual* (4th edn.). Available at: https://www. who.int/publications/i/item/9789240011311 (Accessed 4 December 2021).

WHO. (2021). Meeting report: WHO consultative meeting on a global guidance framework to harness the responsible use of life sciences. Available at: https://www.who.int/publications/i/item/who-consultative-meeting-on-a-global-guidance-framework-to-harness-the-responsible-use-of-life-sciences (Accessed 4 December 2021).

Wintle, B. C., *et al.* (2017). Point of view: A transatlantic perspective on 20 emerging issues in biological engineering. *eLife*, 6, e30247. doi: 10.7554/ eLife.30247.

WOAH. (2015). Biological threat reduction strategy: Strengthening global biological security. Available at: https://www.woah.int/app/uploads/2021/03/en-final-biothreat-reduction-strategy-oct2015.pdf (Accessed 4 December 2021).

WOAH. (2019). Guidelines for responsible conduct in veterinary research: Identifying, assessing and managing dual use. Available at: https://www. woah.int/app/uploads/2021/03/a-guidelines-veterinary-research.pdf (Accessed 4 December 2021).

WOAH. (2021a). Terrestrial animal health code. Available at: https://www.woah. int/en/what-we-do/standards/codes-and-manuals/terrestrial-code-online-access/ (Accessed 4 December 2021).

WOAH. (2021b). Manual of diagnostic tests and vaccines for terrestrial animals. Available at: https://www.woah.int/en/what-we-do/standards/codes-and-manuals/terrestrial-manual-online-access/ (Accessed 4 December 2021).

https://doi.org/10.1142/9781800614826_0007

Chapter 7

A Decade of Responsible Innovation by the AI Community 2012–2022: Analysing Recent Achievements and Future Prospects[*]

Haydn Belfield

The best thing AI researchers can do is vote with their feet, not work with companies that have outcomes you don't agree with. There aren't enough researchers to go around, and attracting enough talent is important, so actually researchers individually have a lot of power. Through soft influence, you can influence a lot.

Demis Hassabis, DeepMind CEO (Tiernan Ray, 2019)

7.1 Introduction

The development and deployment of AI are likely to have important consequences for the global economy, society, and politics. AI deployment is already influencing markets and the economy, justice and distributive decisions by governments, and our elections (Dafoe, 2018; Hajian *et al.*, 2016). In coming years, we may see the use of AI systems with novel

---

[*]This chapter expands upon a version which was first presented as an AIES conference paper (Belfield, 2020).

capabilities in military domains. Increasing use of ever more powerful yet brittle systems carries accident risks, misuse risks, structural risks (Zwetsloot and Dafoe, 2019), risks to human security, and risks of societal inequality and discrimination (Brundage *et al.*, 2018; Koene, 2017).

The AI community has responded to these urgent issues by engaging in activism in order to promote the positive (in their view) societal and ethical effects of AI, and reduce the negative effects. AI community activism may profoundly shape the development and deployment of this important set of technologies — and therefore shape our global economy, society, and politics.

'Activism' includes a broad range of different kinds of social and political campaigning, organising, and advocacy. This encompasses issue framing, agenda setting, standard setting, private discussions with decision-makers, public campaigning in traditional and social media, establishing new fields and organisations, submissions to governmental inquiries, and classic labour tactics of boycotts and strikes. The 'AI community' includes researchers, research engineers, faculty, graduate students, non-governmental organisation (NGO) workers, campaigners and some technology workers, more generally those who would self-describe as working 'on', 'with', and 'in' AI, and those analysing or campaigning on the effects of AI. This paper focuses especially on the AI community within corporate and academic labs in the US and Europe.

This activism has had some notable consequences so far: informing international negotiations, changing corporate strategy, and spurring the growth of research fields (see Section 7.2). Such activism may shape the manner and extent to which AI is militarised, and how AI companies address ethics and safety concerns. The AI community is an important autonomous actor with a distinctive set of viewpoints and interests. It needs to be accounted for in strategic or academic analysis and negotiated with by other actors.

This activism is therefore an important phenomenon in need of theoretical analysis. To date, this analysis has been limited, with much of the discussion in 'grey literature' or in the media (see, e.g., Crawford *et al.*, 2019; Frederick, 2019; Tiku, 2019; Vignard, 2018). We can identify (at least) two important questions for researchers:

- How can we explain the activism by the AI community of recent years?
- What are the future prospects of activism by the AI community?

I have observed and participated in this activism. In 2016, I was part of a team that recommended the UK Government establish the first permanent national advisory body for AI, which became the Centre for Data Ethics and Innovation (CDEI) (Global Priorities Project, 2016). For the past five years — half of the decade which I analyse — I have been part of a team, 'AI: Futures and Responsibility', formed by researchers at the University of Cambridge sister centres, the Centre for the Study of Existential Risk (CSER) and the Leverhulme Centre for the Future of Intelligence (LCFI). In August–September 2019, I was embedded at a leading AI company in San Francisco. This was mainly in order to serve as a lead author on *Toward Trustworthy AI Development* (Brundage *et al.*, 2020). It also gave me the opportunity, as a participant observer, to follow at close hand how employees conceived of and approached responsible innovation in their own practice. From the UN negotiating halls of Geneva to start-up offices in San Francisco, from interviewing whistle-blowers to advising governments and leading technology companies, I have been privileged to have had a ringside seat — and to frequently get into the ring myself.

AI has become one of the key risks that the field of global catastrophic risk (GCR) is concerned about, alongside pandemics, nuclear war, and climate change (Ord, 2020). Its inclusion in this top tier has largely been driven by activism by the AI community. CSER researchers and the broader GCR community have been involved with each of the four examples I discuss below. This illustrates this volume's wider point about GCR researchers' own practice aligning with components of responsible research and innovation.

In this chapter, I overview some examples of recent activism by the AI community. I then apply three different analytical frameworks: epistemic communities, worker organising, and veto players. I end by discussing some key common factors and identifying research questions that could clarify the future prospects of activism by the AI community.

7.2   Recent Examples: 2012–2022

This section is a brief overview of some examples of activism by the AI community over the last decade. The purpose of this section is to ground the later analytical discussion in concrete examples. See AI NOW's 'Year in Review' reports (Campolo *et al.*, 2017, 2019; Crawford and Whittaker, 2016; Whittaker *et al.*, 2018) for a more comprehensive descriptive treatment.

Over the past decade, AI has become a major focus of attention from companies, civil society, and states. This 'AI summer' is often dated from Alexnet (Krizhevsky *et al.*, 2012) — a landmark AI system that achieved ground-breaking results in computer vision/image recognition. This has been followed by substantial progress in game playing, recommender systems, language models, and robotics (AI Index, 2021). Multinational 'tech giants' have invested heavily, and venture capital investment has increased from $3bn to $75bn (OECD, 2021). Governments have announced 700 AI policy initiatives from 60 countries (OECD.AI Policy Observatory, 2022).

The last decade has also seen several high-profile political actions around responsible innovation taken by the AI community. I discuss four main examples: lethal autonomous weapons systems (LAWS), ethics and safety, employee organising, and the beginning of regulation of the AI sector. These four do not cover the entire range of activism, but are intended to be broadly illustrative. Under these headings, I proceed in a roughly chronological manner.

7.2.1 *Lethal Autonomous Weapons Systems (LAWS)*

LAWS are weapons systems that can autonomously select and kill targets (International Committee of the Red Cross, 2021). The AI community has been deeply involved with the debate over an international ban on LAWS (Russell *et al.*, 2018). Not only is the AI community a relevant expert community for advising on this debate but it is also being directly asked to work on the research and development (R&D) of LAWS (Verbruggen, 2019).

In early 2013, the Campaign to Stop Killer Robots was formally launched to promote an international ban on the development and use of LAWS. This was a key moment of epistemic community formation and a focus for worker organising.[1] In late 2013, the Convention on Certain Conventional Weapons (CCW) agreed to begin considering LAWS. The CCW was the forum for negotiations over banning cluster munitions, landmines, and blinding laser weapons. The AI community played key roles in the adoption of LAWS as an issue by arms control NGOs, the establishment of the UN process, and the ongoing work of the CCW (Bahçecik, 2019). Researchers remain 'overwhelmingly opposed' to working on LAWS (Zhang *et al.*, 2021).

---

[1]The campaign's roots stretch back at least to 2004 (Carpenter, 2014).

The Campaign includes many members of the AI community. Their activism ranges from personal discussions with diplomats at the CCW to mass media 'viral videos' (*Slaughterbots*, 2017; *Slaughterbots — if human: kill() — YouTube*, 2021). A key tactic has been the organisation of mass open letters. In July 2015, an open letter on LAWS expressed community concern. It has been signed to date by 4,500 AI and robotics researchers (Future of Life Institute, 2016). This was followed by another open letter on LAWS (Future of Life Institute, 2017a), when the Meeting of Experts changed to a Group of Governmental Experts (GGE) at the CCW.

This effort has also included intra-community organising. One example is the April 2018 Korea Advanced Institute of Science and Technology (KAIST) boycott. The 50 signatories committed to boycotting all collaborations with any part of KAIST, due to concerns that a KAIST centre had an LAWS collaboration with Hanwha Systems, a leading South Korean arms company. KAIST soon clarified that they would not work on LAWS (Future of Life Institute, 2018). Activism has also included extensive and intense negotiations within technology companies, as discussed as follows.

Negotiations within the CCW GGE process are ongoing. There has been useful progress made: 'Guiding Principles' have been agreed upon, with elaborations of how this could be operationalised as a code of conduct and a set of guidelines (Moyes, 2020). Militaries are acknowledging the challenges of weapons reviews and Test & Evaluation, Validation & Verification (TEVV) for AI systems, and are exploring and sharing best practice (Flournoy *et al.*, 2020). However, prospects for a binding treaty being agreed within this consensus-based forum are slim. Campaigners and states may establish a new treaty outside the CCW process, as they did for land mines and cluster munitions.

Scepticism has been expressed about the coherence of the epistemic community on military AI, especially about whether it is presently sufficient to play a similar role to the ABM community (Maas, 2019; Payne, 2018; Rosert and Sauer, 2021). It clearly has not yet been sufficient to pass an international ban or block military AI R&D efforts. However, it has put LAWS firmly on the international agenda, and the Campaign's prospects look similar to the campaigns on the landmines and cluster munitions at similar stages. For now, it has lastingly complicated the relationship between the US military and US tech firms (most importantly Google). Moreover, we are considering activism on a wider set of topics than just LAWS. The AI community's policy projects also include emphasising ethics, societal benefit and safety, and opposing particular choices by company management.

7.2.2 *Ethics and safety*

There has been sustained activism from the AI community to emphasise that AI should be developed and deployed 'responsibly' — in a safe and beneficial manner (Askell *et al.*, 2019). This has involved open letters, AI principles, the establishment of new centres, and influencing governments.

The Puerto Rico Conference in January 2015 was a landmark event to promote the beneficial and safe development of AI. It led to an open letter signed by over 8,000 people, including many prominent AI researchers, calling for 'expanded research aimed at ensuring that increasingly capable AI systems are robust and beneficial', and a research agenda to that end (Future of Life Institute, 2015).

The Asilomar Conference in January 2017 led to the Asilomar AI Principles, signed by several thousand AI researchers (Future of Life Institute, 2017b). Over 80 sets of principles from a range of groups followed (Whittlestone *et al.*, 2019).

The AI community has established several research groups to understand and shape the societal impact of AI. AI conferences have also expanded their work to consider the impact of AI. New groups include (but are by no means limited to):

- Fairness, Accountability, and Transparency in Machine Learning (FAT ML) (December 2014)
- OpenAI (December 2015)[2]
- Centre for Human-Compatible AI, UC Berkeley (August 2016)
- Leverhulme Centre for the Future of Intelligence (October 2016)
- Algorithmic Justice League (November 2016)
- DeepMind Ethics and Society (October 2017)
- AI Now Institute (November 2017)
- UK Government's Centre for Data Ethics and Innovation (November 2017)[3]

---

[2] OpenAI was a non-profit and is now a company with a non-profit mission.

[3] The Centre was proposed by and advocated for by the AI community, and analyses AI's societal and ethical implications.

- The Ada Lovelace Institute (March 2018)
- Institute for Human-Centered Artificial Intelligence, Stanford (March 2019)
- China–UK Research Centre for AI Ethics and Governance (November 2019)
- Centre for Security and Emerging Technology (CSET), Georgetown University (January 2019)
- Institute for Ethics in AI, Oxford University (February 2021)

Especially notable is the Partnership on Artificial Intelligence to Benefit People and Society (September 2016). The Partnership is the leading non-profit coalition of AI technologists, bringing together the world's leading AI companies (and over 90 companies and non-profits) to explore best-practice recommendations for the community as a whole (Partnership on AI to Benefit People & Society, 2016).

Within existing organisations, new dedicated AI alignment, and governance teams have been set up at the Montreal Institute for Learning Algorithms (MILA) and leading AI companies OpenAI, DeepMind, and Google Brain. It is important that this community not just be focused on Europe and the US but also include global perspectives, especially from the Global South and East Asia, where much of AI progress is occurring.

The academic and practitioner fields of technical AI alignment (or safety) and non-technical AI governance (including ethics and policy) have become firmly established. For example, there are several dedicated workshops on AI alignment at leading AI conferences. These regular workshops embed alignment in the wider field and provide a publication venue for high-quality AI alignment research.

These new centres and associated field-building work are breaking new research ground and having policy impact. This can be seen in two major recent publications, of which I had an intimate view as one of the authors.

'The malicious use of artificial intelligence: Forecasting, prevention, and mitigation' was the first major publication to examine emerging risks at the intersection of artificial intelligence, cybersecurity, physical security, and information manipulation (Brundage *et al.*, 2018). It was a collaborative report written by 26 co-authors from 14 institutions, based on several workshops co-organised by the Centre for the Study of Existential Risk and Oxford's Future of Humanity Institute in 2017. It received a warm reception from policymakers, academics, and technologists — including a UK Minister, the Commander of the Australian Defence College, and the former president of the Association for the Advancement of Artificial Intelligence (AAAI). The report has become a landmark publication, currently cited over 500 times, including by several national strategies and reports from the UK House of Lords and the US Department of Defense.

'Malicious Use' began a widespread conversation about AI publication and release norms, which is changing the behaviour of AI research organisations. The report encourages greater care and patience in the publication of results, and suggests that pre-publication risk assessments may be necessary in some circumstances. In early 2019, OpenAI, a major AI research company, carried out a pre-publication risk assessment and adopted a policy of 'staged release' for their potentially dual-use AI language model, GPT-2 (Solaiman *et al.*, 2019). OpenAI did not immediately release the full model, to allow time to examine the possibility of its misuse in increase text-based disinformation (e.g., news articles or spam emails). This decision was influenced by several of its researchers being co-authors on 'Malicious Use'. This decision started a widespread debate — and behaviour change — across the ML community. The Partnership on AI subsequently developed a more detailed report on AI Publication Norms for use by AI research companies (Campbell, 2021).

The team behind 'Malicious Use' collaborated again on 2020's *Toward Trustworthy AI Development: Mechanisms for Supporting Verifiable Claims*. It was written with 59 international co-authors from 29 groups including leading technology companies. It proposes ten immediate, concrete mechanisms AI developers can take to make more verifiable claims in three areas: institutional, software, and hardware. Several of these mechanisms are now being explored and implemented by the Partnership on AI and several technology companies (Avin *et al.*, 2021; Hua and Belfield, 2021).

7.2.3 *Organising*

2018 saw an upturn in political organising, especially within large technology companies: the 'tech resistance'.

The most prominent example of this activism has been Google. Google's involvement with Department of Defense's (DoD) Project Maven was revealed in March 2018. Project Maven focused on drone image recognition, and was widely seen as a precursor to work on LAWS. 3,000 employees signed an open letter opposing it in April 2018 (Google Employees, 2018). In June 2018, Google announced it would not renew the Project Maven contract and released its AI Principles. In October 2018, Google also dropped out of the DoD's JEDI cloud bidding process.

However, in August 2018, Google's secret 'Project Dragonfly' (a censored Chinese search engine) was revealed. This was opposed by another open letter. Two anonymous employees wrote in an email circulating the Dragonfly letter: 'Individual employees organizing against the latest dubious project cannot be our only safeguard against unethical decisions. This amounts to unsustainable ethics whack-a-mole' (Google Employees Against Dragonfly, 2018).

In November 2018, 20,000 employees and contractors took part in a one-day strike, or 'Google walkout', protesting sexual harassment and misconduct. This was also part of the broader #MeToo movement. This walkout contributed to the end of 'forced arbitration' for full-time workers (Google Walkout for Real Change, 2019).

This activism also occurred at other leading Big Tech companies. In June 2018, in a reaction to the Trump administration's family separation policy, several employee groups released open letters challenging corporate partnerships with Immigration and Customs Enforcement (ICE) and Customs and Border Protection (CBP) (Amazon Employees, 2018; Microsoft Employees, 2018; Salesforce Employees, 2018).

However, over subsequent years, there has been a backlash to this organising. Several employees who played leading roles have left, often under conditions of pressure or outright firing, such as Timnit Gebru and Emily Cunningham. Some employees have felt stymied by internal organising and have quit to become whistle-blowers, such as Frances Haugen and Ifeoma Ozoma. Key internal forums, like Google's TGIF all-hands meeting, have been weakened or ended.

Due in part to dissatisfaction with the limits of corporate self-governance, internal organising, and 'ethics whack-a-mole', the focus of many internal organisers has now turned to regulation. This trend is demonstrated by Meredith Whittaker's career. Whilst a Google employee, she co-founded AI NOW and was a core organiser of the Google Walkout. She then left, or was forced out under pressure, and was then an advisor to Lina Khan, Chair of the Federal Trade Commission (FTC), who is attempting to rein in Big Tech and break up Meta. Meredith Whittaker is now President of Signal.

7.2.4 *Regulation*

The past decade has seen steady progress towards regulation of the civilian applications of AI (military applications were discussed in Section 7.2.1). This began with a proliferation of principles and strategies. These were seen as insufficient to the societal challenges posed by AI, especially given ongoing scandals over this period. Regulation has been drafted and will soon be passed in several jurisdictions, most notably the EU. Over the coming years, this regulation will be operationalised into detailed technical standards, through the EU's European Committee for Standardization and the European Committee for Electrotechnical Standardization (CEN-CENELEC), and the US' National Institute of Standards and Technology (NIST). Enforcement will then follow — and will have significant teeth.

The EU AI Act is the world's first — and most significant — move to regulate AI systems as such. It bans some forms of AI systems such as 'social scoring' and puts certain requirements, such as transparency, on all providers of AI systems. It also identifies eight application areas, such as education, immigration, welfare, and policing, as 'high risk' and requires that AI systems used in these areas be tested to conform to mandatory requirements, such as around robustness and accuracy or human oversight.

This may have substantial influence over AI regulation worldwide through the 'Brussels Effect'. The EU is such a large, rich, and advanced economy and jurisdiction. Other states may copy it (the *de jure* effect), and companies may adopt the EU's stringent standards across their products and services worldwide (the *de facto* effect) (Siegmann and Anderljung, 2022).

The EU has passed the Digital Markets Act (DMA) and Digital Services Act (DSA), which place additional responsibilities and codes of conduct on particular 'gatekeeper' companies, that is, Big Tech. This shift to an *ex ante* 'gatekeeper' regime of additional scrutiny under antitrust/competition law is being replicated in the UK. In the US, the FTC and several state attorneys general are also bringing big antitrust cases against Big Tech. The EU, USA, UK, China, Taiwan, and others are also providing significant funding and introducing new regulation to promote their AI hardware ecosystems, from semiconductor manufacturing to cloud computing.

AI is also firmly on the agenda of various UN agencies and international organisations. The Organisation for Economic Co-operation and

Development (OECD) and the World Economic Forum (WEF) have established organisations on AI policy, and the Global Partnership on Artificial Intelligence (GPAI) was created in 2020. The International Telecommunication Union's (ITU) AI for the Good Summit series has become the leading United Nations platform for dialogue on AI. Finally, the Secretary-General established a Panel on Digital Cooperation which Melinda Gates and Jack Ma co-chair. Finally, the Council of Europe's Ad hoc Committee on Artificial Intelligence (CAHAI) recommended a legally binding treaty, covering impact assessments and risk classifications.

The AI community has been intimately and extensively involved with this shift to regulation. It has vocally proposed, contributed to, and advocated for regulation.

For example, the EU AI Act is largely based on work by the AI community. The High-Level Expert Group on Artificial Intelligence (AI HLEG) to the European Commission was made up of 52 experts from academia, civil society, and industry. Their work led to Ethics Guidelines and policy and investment recommendations for trustworthy AI (AI HLEG, 2019). Those guidelines included seven requirements, such as technical robustness and safety, that AI systems should meet, which are now in the Act. Their role is formalised as an 'AI Board' in the Act. The AI community also contributed through feedback submissions throughout the drafting process and participation in the 'AI Assembly'.

The AI community has taken up influential roles in government and regulators. For example, in the USA, as previously noted, Khan and Whittaker have moved from being critics to regulators. Other prominent appointments include the founding Director of CSET, Jason Matheney, taking up an influential position in the White House's Office of Science and Technology Policy (OSTP) and the National Security Council (NSC).

Partly, this reflects churn in US politics. The US is unusual in how many appointments across the bureaucracy are politically appointed — it is usual for a change in administration to be accompanied by a host of people from companies and think-tanks moving into government. Many of those involved in debates on LAWS, ethics and safety, and organising lean Democratic so it is not surprising they have joined the Biden Administration. However, I would suggest this also reflects the maturation of the AI community and the growing sense that regulation is now required for responsible innovation.

7.3 Analytical Frameworks: Three Lenses

I reviewed four examples of activism over the last decade, including some key successes. I now draw on three different analytical frameworks: epistemic communities, worker organising, and veto players. These can provide insights into why this activism is occurring, how these successes have been achieved, and what its future prospects are.

I address each framework in turn. For each framework, I briefly describe it and explain its relevance to our discussion. I then discuss what factors are seen as predictive of success for the relevant group within the framework, and ask to what extent those factors apply to our case — and will continue to apply over the next few years.

7.3.1 *Epistemic communities*

An epistemic community is a network of knowledge-based experts: 'professionals with recognised expertise and competence in a particular domain and an authoritative claim to policy-relevant knowledge within that domain' (Haas, 1992). Crucially, they also share causal and principled beliefs about their domain, notions of validity, and a common policy project (Adler, 1992). Historical examples include nuclear physicists, chemists, biologists, and climate scientists. Epistemic communities have been identified as playing key roles in several policy debates, such as the role of the nuclear weapons scientists and strategists in the 1972 Anti-Ballistic Missile (ABM) Treaty.

This framework is relevant to the AI community. These communities are technical experts who have a clearly recognised and valid claim to authority and expertise in the domain of AI. There is a strong set of shared causal beliefs and notions of validity, and common policy projects.

In general, when are epistemic communities more likely to be persuasive? Mai'a K. Davis Cross identifies five key factors: (1) the issue is uncertain and salient; (2) the community has access to, and understanding of, decision-makers and other actors; (3) there is 'policy field coherence'; (4) they seek to influence an early or technocratic phase in the policy process; and (5) they are seen as credible and have a more cohesive, certain coalition than their competitors (Cross, 2012).

I suggest that these factors currently apply to the AI community, and are likely to largely continue to apply.

(1) The ethical, military, and societal implications of AI are a complex and new set of issues. It has become politically salient over the last decade.

It is likely to remain quite uncertain. The extent to which there is a sense of perceived crisis may fade over time as the issue becomes less new and begins to be untangled. However, it is likely to remain technically complex, as well as politically salient, as AI continues to impact our economy and society.

(2) The AI community has access to top decision-makers, both corporate and political. They have also been able to anticipate other actors' preferences and actions, such as anticipating those of the different national delegations during the international LAWS negotiations.

This high-level access is likely to continue — there is no sign of policymaker interest slacking. It is open whether they will continue to be anticipatory.

(3) 'Policy field coherence' refers to a field having respected quantitative data, the issue involving technical systems, and the norms and goals of the community being compatible with existing institutional norms. The issues for the AI community have involved the interaction of human-made technical systems with social systems, instead of just involving social systems, and there has been respected quantitative data on, for example, the extent of bias or the frequency of mistakes by ML systems (Koene, 2017).

There will likely continue to be policy field coherence. Indeed, the availability of quantitative data is likely to improve over time. However, the lack of quantitative data about the humanitarian impact of LAWS (which will continue until they are deployed) may continue to limit the persuasiveness of those pushing for an international ban.

(4) The AI community has sought to influence the initial terms of the debate and to focus on subsystem, technocratic phases. For example, activism in 2015 helped shape the terms of the debate around ethical, safe,

and beneficial AI — and then that was translated into national ethical principles and strategies.

Over time, the terms of the debates around the development and deployment of AI are likely to become more fixed, and activism will have to seek to influence later stages in decision-making processes. The societal implications of AI may also become more pronounced, and/or entangled with broader political beliefs. However, ongoing technical breakthroughs may create continued capability jumps (such as language models and fake news generation), routinely posing new policy problems.

(5) The AI community has been quite coherent and certain of its aims in some respects: pushing for serious consideration of the military and societal impacts of AI by companies and governments. It seems more cohesive than those pushing to deproblematise the issue. It also shares a high level of professional norms and status.

It is unclear whether the community will continue to be more coherent and certain than competing networks, especially in the debate over the acceptability of working on military AI, including LAWS. The network pushing for an accommodation with militaries may prove more cohesive and certain. Community credibility seems unlikely to radically change. Marginal changes could occur, for example, through further democratisation of ML through Massive Open Online Courses (MOOCs) and other online platforms.

Overall then, the AI community has achieved some successes as an epistemic community as the scope conditions (1), political opportunity structure (2), and policy field coherence (3) have been favourable, and the AI community has sought to influence early and technocratic phases in the policy process (4) and built fairly cohesive coalitions (5).

Factors likely to continue include high-level access being good, being able to deal with technocratic aspects of decision-making, sharing a high status, and being able to draw on evidence on which they are experts. However, over the next few years, the novelty will decrease, the terms of debate will become more set, and the entire issue/area may become more politicised. The window of influence may be shrinking, but it is unlikely to close, as the issue/area will remain uncertain and complex. A key question for further research is whether the AI community will continue to be cohesive and certain in its aims.

7.3.2 *Worker organising and bargaining*

Freeman and Medoff influentially distinguish between the 'two faces' of organised labour (Freeman and Medoff, 1984). The first, 'monopoly face', is that of union monopoly power used to raise members' wages. Now, however, the emphasis is typically more on protecting wages, by offsetting informational and power asymmetries between workers and management (TUC, 2019). More relevant to our focus, however, is their second, 'collective voice/institutional response' face. This builds on Hirschman's conception of 'voice' as the ability to 'change, rather than escape from (i.e., exit), an objectionable state of affairs' (Hirschman, 1970).

Worker organising and bargaining means their ability 'to "voice" their concerns and demands rather than immediately "exit" — that is, quit the job', 'to voice complaints and see them addressed through collective bargaining' (Freeman and Medoff, 1984). Worker or employee voice — whether formally through a union or not — has become a key topic for disciplines such as labour economics, industrial relations, and organisational behaviour (Wilkinson *et al.*, 2020).

It need not refer just to improving working conditions but also to encouraging one's firm to advocate for public policies. For example, US corporate engagement on LGBT rights has been largely driven by employee organisations in highly educated workforces advocating for management to take public stands for LGBT rights (Drewry and Maks-Solomon, 2019).

Key to this framework is the ability of different actors to have an agreement on their terms, that is, the 'bargaining power' of workers and management, based on the 'ability to impose costs on the other side for failing to agree and to avoid or absorb its own costs from failing to agree' (Dau-Schmidt and Ellis, 2011).

This framework is relevant to the AI community. Almost all members of the AI community are employees, rather than business-owners. Indeed, the AI community is largely located within fairly large corporate and academic labs.

Formal unionisation rates are low in technology companies, though higher in academia (Fanti and Buccella, 2019). However, it is clear that the AI community has substantial bargaining power, of the type described in our opening quote. This is reflected in high salaries. *The New York*

Times reports that 'A.I. specialists with little or no industry experience can make between $300,000 and $500,000 a year in salary and stock. Top names can receive […] millions' (Metz, 2018; see also Nolan and Coulter, 2022). Nevertheless, groups like the Tech Workers Coalition have emphasised the difference in power across the industry — from a researcher with a ML PhD to a gig economy Mechanical Turk contractor. The Coalition's work began in attempting to organise across more of the tech workforce (Upadhya, 2018). The organising examples described above often involved workers adapting classic collective bargaining tactics for their situations.

While this framework is directly relevant to employees, more generally, AI community activism can be framed as a set of strategic interactions between actors, each with different incentives, resources, and constraints — where outcomes are the result of bargaining between these actors (Leung, 2019).

In general, when are workers more likely to be persuasive (or successful) with the management of their organisation? Dau-Schmidt and Ellis identify five key economic factors: (1) the nature of the organisation's product and services; (2) the structure of bargaining; (3) the organisation's technology of production; (4) general economic conditions; and (5) the employees' commitment to collective action (Dau-Schmidt and Ellis, 2011).

I suggest that these factors currently benefit the AI community, and are likely to largely continue to benefit it.

(1) Corporate organisations' products and services typically require ongoing maintenance, so management is less able to absorb costs from failing to agree with its employees. Also, large tech companies are generally consumer-facing (B2C) rather than business-facing (B2B) firms, which makes them more susceptible to public, i.e., consumer, sentiment. Google is more susceptible than Palantir to consumer sentiment because it relies on the public using its services, rather than Palantir which relies more on large clients such as governments. Tech workers have been adept at mobilising media interest and public support. The nature of these products and services is unlikely to change dramatically in the coming years.

(2) The structural landscape of bargaining is largely tilted in favour of employers. While many tech workers are well paid enough to have a decent amount of runway, and are able to 'exit' to similarly well-paid jobs

elsewhere, they are hampered by the lack of unionisation. Another issue is the widespread use by tech companies of non-disclosure agreements (NDAs). This tends to limit workers' ability to publicly voice concerns — even after they leave the company.

This is likely to continue, but may diminish marginally if there are explicit offers of financial or legal support from worker groups like the Tech Worker's Coalition or civil society groups like the American Civil Liberties Union (ACLU). Moreover, several centre-left politicians in the US and Europe propose more antitrust scrutiny of tech companies, and to change union regulation to empower them, which would shift the landscape. On the other hand, if AI grows in geopolitical importance and is increasingly seen as key to economic competitiveness and state security, states may constrain researchers and employees through new legislation — as in nuclear power (Leung, 2019).

(3) Organisations employing members of the AI community typically depend on highly skilled workers. Replacements are hard to obtain, and the organisation cannot continue with just a skeleton crew. This means that the organisations are less able to absorb the costs of failing to agree to a deal, improving worker bargaining power.

This is likely to continue, but may change marginally. If these organisations turn from more R&D-style work to more implementation-style and commercial work, then workers may not need to be as highly skilled. This may shift bargaining power.

(4) A key question about the general economic conditions is the balance of supply and demand of potential replacement workers. Estimates of the 'talent pool' vary widely, from 22,400–36,500 (Gagné, 2019, 2020) to 300,000 (TenCent Research Institute, 2017). The high salaries commanded in the field demonstrate that supply is not meeting demand.

However, various sources of evidence point to an increase in the talent supply in coming years (AI Index 2021; AI Index Steering Committee, 2022). Several governments have made large commitments of money for education: the UK is supporting 1,000 PhDs (Department for Business, Enterprise & Industrial Strategy, 2019). The CRA Taulbee Surveys show continued increase in the number of degrees and enrolment, at doctorate, postgraduate, and undergraduate levels (Zweben and Bizot, 2019, 2020a, 2020b, 2022), though this is limited by faculty shortages (Kinoshita, 2022; Zwetsloot and Corrigan, 2022). Note that this does not include quasi-academic routes: MOOCs such as Andrew Ng's or fast.ai and coder bootcamps.

We should expect an increase in the supply of talent over the coming years. A PhD in ML, for example, typically takes four years to complete in the UK and six years in the USA. If someone was inspired by AlphaGo in 2016, they might only have joined the workforce in 2022. Other things being equal, this increase in the number of potential workers would tend to decrease worker power.

However, this assumes that supply increases more than demand. The following years may well see demand for ML talent continue to increase — keeping track with or even exceeding the increase in supply. This demand may increase as AI and robotics become more technologically ready or industrialised. If data and computing power continue to grow, more areas may become well suited to ML approaches, which might also increase demand. The future balance of supply and demand is a key question.

(5) Commitment to collective action is a similar factor to epistemic community cohesion. Willingness and ability to (threaten) exit is high (Zhang *et al.*, 2021). There have been several high-profile exits over 'employee voice' issues (Nolan, 2018; O'Sullivan, 2019; Poulson, 2019). A small-scale survey of UK tech workers suggests that 16% of all people in AI have left their company over the issue of working on products they felt might be harmful for society, compared to 5% of all tech workers (Miller and Coldicutt, 2019).

A key question is whether the predicted new 'talent' will be socialised to a similar extent, and through similar means, to existing workers. A larger AI community, which has not all been to the same small academic conferences for several years, may be less cohesive. Conversely, enrolment in university 'ethics in AI' courses may make the next generation more ethically engaged. Also, if the momentum behind the political organising continues, it will increasingly be seen as the norm. Key organisers might receive more training and support. The ad hoc arrangements that have characterised the political organising of the AI community so far might become more institutionalised.

Overall then, the AI community has achieved some success as workers organising and bargaining with their employers. This may be attributed to the organisational products and services (1), organisational production technology (3), and the general economic conditions (4) all being favourable — though the structure of bargaining has not been (2) — and the AI community having been fairly committed to collective action (5).

Factors likely to continue include the products and services being consumer-facing and difficult to stockpile, reliance on high-skilled labour, and the unequal bargaining structure. However, it is unclear what the balance of talent supply and demand will be, and to what extent the AI community will continue to be committed to collective action. These are key questions for further research.

7.3.3 *Veto players*

I will briefly explore a third and final analytical lens: veto players. Veto players are individual or collective actors whose agreement (by majority rule for collective actors) is required for a change of the status quo (Tsebelis, 2002). This framework is generally applied to comparative politics, where it has been influential (Jahn, 2011) though contested (Ganghof, 2003). Notable examples include comparing political decision-making in parliamentary and presidential systems (Tsebelis, 1995). It is also relevant to the AI community and can shed some novel, interesting light on our discussion.

The debate in the USA over the militarisation of AI and LAWS can be viewed as the interaction of three actors: employees, management of AI organisations (academic groups or technology companies), and the government (Leung, 2019). Crucially, one can view this as a veto player situation. The current status quo is one in which the AI community does not generally research LAWS. There are few US AI organisations, especially the most prominent, engaged in LAWS R&D for the US government. The US government, and management at several organisations, would like that to change — to a situation in which there is widespread collaboration from many organisations, including the most prominent. For the status quo to change, all actors must agree. However, a significant group of employees is not agreeing to this change.

In general, when are veto players more likely to be able to prevent a change of the status quo? Tsebelis identifies three key factors: (1) the number of veto players; (2) lack of congruence; and (3) cohesion (Tsebelis, 2002). I suggest that these factors currently apply to some extent to the AI community, and are likely to largely continue to apply.

(1) The number of veto players is not particularly high. It seems that perhaps management and the government believed they were in a two-player

game. They seem unprepared for the activism of the AI employees on this topic, which created a three-player game with three veto players.

(2) The policy positions of the players have been rather dissimilar. Many governments and company managers want collaboration on military AI, and specifically LAWS, to not only be widespread but also unproblematised. Many, perhaps most, employees do not.

This seems likely to continue, though; for example, the process of developing and exploring the proposed DoD ethics principles (Defense Innovation Board, 2019) may achieve an accommodation between Silicon Valley and the Pentagon, increasing congruence between the players.

(3) The relative similarity of policy positions amongst the AI community has been discussed above. There is clearly a debate about the extent to which they should be involved in the militarisation of AI, especially LAWS R&D, and whether there should be an international ban. This debate is likely to continue.

The AI community has achieved some successes as a veto player as the congruence between the players has been low (2), and the AI community has been fairly coherent (3). This situation is broadly likely to continue. However, whether the AI community will continue to be cohesive is a key question.

7.4 Discussion and Further Research

What are the prospects for the AI community to be able to successfully continue its activism over the coming years? The above analysis indicated two key areas in which further research is warranted: the balance of talent supply and demand and the cohesiveness of the AI community.

I identified a small, narrow talent pool as a key factor in the bargaining power of the AI community. We need to be able to assess relative supply and demand of AI talent over the coming years. Further quantitative research is needed into, for example, the current size of the talent supply, collated salary information, enrolment rates, or future demand projections for AI talent.

The AI community has several resources at its disposal to maintain and deepen its social, organisational, ethical, and political cohesion. The AI community has a strong shared culture, with strong norms of responsibility, 'do-it-yourself', and mutual support. The community publishes at and attends the same major conferences (such as AAAI, NeurIPS, ICML and IJCAI), publishes on the same sites such as arXiv and GitHub, and uses the same libraries such as TensorFlow or PyTorch. Virtually none of these outlets charge for publication or access, and are often maintained by the community. This culture can be seen in the closed-access journal boycott (Dietterich, 2018) and the name change from NIPS to NeurIPS (Anandkumar, 2018).

The AI community has made use of institutional enabling structures: the Future of Life Institute as a coordinator of international open letters; the Tech Workers Coalition and the Partnership on AI as distributors of best practice; corporate digital tools such as email mailing lists and internal chat rooms; and social media such as X (previously, Twitter) and Medium as a way of communicating demands. These structures could be developed, better funded, and institutionalised — and key organisers could receive more training and support. Further research into the AI community's ability and willingness to maintain and deepen its cohesion — and the structures and institutions that support that — is needed.

Further research could also extend our analysis in many ways. It could extend to a broader range of regional and sectoral contexts, for example, authoritarian countries such as China or Russia, or particular industries such as industrial robotics. It could extend to a broader range of actors with influence over companies, such as investors and divestment campaigns. And it could extend to other analytical frameworks. For example, there may be a trend across general-purpose technologies (GPTs), such as cryptography and biotechnology, for researcher and employee power to decrease compared to companies and governments over time (Leung, 2019). To what extent this more general trend applies to AI and over what timescale (years or decades) are key questions for further research.

7.5 Conclusion

The AI community is acting together — it is organised. It has turned this attention to activism on responsible innovation. It has won some key successes. And yet its future prospects are uncertain. Useful insights can be drawn from the literature on epistemic communities and worker organising. There are good precedents for highly skilled groups engaging in activism as epistemic communities and valuable employees, thereby influencing policy and corporate outcomes. Indeed, the AI community has already had some clear successes around LAWS, ethics and safety, and employee organising. Whether this will continue is less clear. The future balance of AI talent supply and demand and the future cohesion of the AI community are key questions for further research.

Acknowledgements

This article builds on material from an earlier open-access article by the author, who is the permanent holder of all rights to the article. Please see Haydn Belfield (2020). Activism by the AI Community: Analysing Recent Achievements and Future Prospects. In *Proceedings of the AAAI/ ACM Conference on AI, Ethics, and Society (AIES '20)*. Association for Computing Machinery, New York, NY, USA, 15–21. https://doi. org/10.1145/3375627.3375814.

References

Adler, E. (1992). The emergence of cooperation: National epistemic communities and the international evolution of the idea of nuclear arms control. *Int. Organ.*, 46(1), 101–45.

AI Index. (2021). AI Index 2021. Stanford institute for human-centered artificial intelligence. Available at: https://hai.stanford.edu/research/ai-index-2021 (Accessed 6 December 2021).

AI Index Steering Committee. (2022). *The AI Index 2022 Annual Report*. Stanford: Stanford Institute for Human-Centered Artificial Intelligence. Available at: https://aiindex.stanford.edu/report/.

Amazon Employees. (2018). I'm an Amazon employee. My company shouldn't sell facial recognition tech to police. *Medium*. Available at: https://medium. com/@amazon_employee/im-an-amazon-employee-my-company- shouldn-t-sell-facial-recognition-tech-to-police-36b5fde934ac (Accessed 6 December 2021).

Anandkumar, A. (2018). #ProtestNIPS change.Org petition. Change.org. Available at: https://www.change.org/p/members-of-nips-board-protestnips- nips-acronym-encourages-sexism-and-is-a-slur-change-the-name (Accessed 6 December 2021).

Askell, A., Brundage, M., and Hadfield, G. (2019). The role of cooperation in responsible AI development. *arXiv:1907.04534 [cs]*. Available at: http:// arxiv.org/abs/1907.04534 (Accessed 2 February 2022).

Avin, S., *et al.* (2021). Filling gaps in trustworthy development of AI. *Science,* 374(6573), 1327–29.

Bahçecik, Ş. O. (2019). Civil society responds to the AWS : Growing activist networks and shifting frames. *Global Policy*, 10(3), 365.

Belfield, H. (2020). Activism by the AI community: Analysing recent achievements and future prospects. In *Proceedings of the AAAI/ACM Conference on AI, Ethics, and Society*, New York, USA: Association for Computing Machinery, pp. 15–21. http://doi.org/10.1145/3375627.3375814.

Brundage, M., Avin, S., Wang, J., Belfield, H., *et al.* (2020). Toward Trustworthy AI Development: Mechanisms for Supporting Verifiable Claims. arXiv:2004.07213.

Campbell, R. (2021). *Managing the Risks of AI Research: Six Recommendations for Responsible Publication*. San Francisco, CA: Partnership on AI. Available at: https://partnershiponai.org/paper/responsible-publication-recommendations/ (Accessed 2 February, 2022).

Campolo, A., Sanfilippo, M., Whittaker, M., and Crawford, K. (2017). *AI Now 2017 Report*. New York: AI Now. Available at: https://ainowinstitute.org/publication/ai-now-2017-report-2 (Accessed 30 January 2024).

Carpenter, C. (2014). *"Lost" Causes: Agenda Vetting in Global Issue Networks and the Shaping of Human Security*. Illustrated edition. Ithaca; London: Cornell University Press.

Crawford, K. and Whittaker, M. (2016). *AI Now 2016 Report*. New York: AI Now. Available at: https://ainowinstitute.org/news/2017-symposium-2 (Accessed 30 January 2024).

Crawford, K., *et al.* (2019). *AI Now 2019 Report*. New York: AI Now. Available at: https://ainowinstitute.org/publication/ai-now-2019-report-2 (Accessed 30 January 2024).

Cross, M. K. D. (2012). Rethinking epistemic communities twenty years later. *Rev. Int. Stud.*, 39(1), 137–160.

Dafoe, A. (2018). *AI Governance: A Research Agenda*. Oxford: Centre for the Governance of AI (GovAI). Available at: https://www.governance.ai/research-paper/agenda (Accessed 2 February, 2022).

Dau-Schmidt, K. G. and Ellis, B. C. (2011). The relative bargaining power of employers and unions in the global information age: A comparative analysis of the United States and Japan.

Defense Innovation Board. (2019). *AI Principles: Recommendations on the Ethical Use of Artificial Intelligence by the Department of Defense*. Washington DC. Available at: https://media.defense.gov/2019/Oct/31/2002204458/-1/-1/0/DIB_AI_PRINCIPLES_PRIMARY_DOCUMENT.PDF (Accessed 6 December, 2021).

Department for Business, Enterprise & Industrial Strategy. (2019). *AI Sector Deal*. London: HM Government. Available at: https://www.gov.uk/government/publications/artificial-intelligence-sector-deal/ai-sector-deal (Accessed 6 December 2021).

Dietterich, T. (2018). Statement on nature machine intelligence. Available at: https://openaccess.engineering.oregonstate.edu/home (Accessed 6 December 2021).

Drewry, J., and Maks-Solomon, C. (2019). Why do corporations engage in activism on LGBT issues? *Acad. Manage. Proc.*, (1), 19253.

Fanti, L. and Buccella, D. (2019). When unionisation is profitable for firms in network industries. *Metroeconomica*, 70(4), 711–22.

Flournoy, M. A., Haines, A., and Chefitz, G. (2020). *Building Trust Through Testing: Adapting DOD's Test & Evaluation, Validation & Verification (TEVV) Enterprise for Machine Learning Systems, Including Deep Learning Systems*. Washington DC: Center for Security and Emerging Technology. Available at: https://cset.georgetown.edu/event/building-trust-through-testing/ (Accessed 2 February 2022).

Frederick, K. (2019). The civilian private sector: Part of a new arms control regime? *The Raisina Files 2019 Debating Future Frameworks in a Disrupted World.* Available at: https://www.orfonline.org/expert-speak/the-civilian-private-sector-part-of-a-new-arms-control-regime-57345/ (Accessed 6 December, 2021).

Freeman, R. and Medoff, J. (1984). *What Do Unions Do*. NY: Basic Books.

Future of Life Institute. (2015). AI open letter: Research priorities for robust and beneficial AI. Available at: https://futureoflife.org/ai-open-letter/ (Accessed 6 December, 2021).

Future of Life Institute. (2016). Autonomous weapons open letter: AI & robotics researchers. Available at: https://futureoflife.org/2016/02/09/open-letter-autonomous-weapons-ai-robotics/ (Accessed 6 December 2021).

Future of Life Institute. (2017a). An open letter to the united nations convention on certain conventional weapons. Available at: https://futureoflife.org/autonomous-weapons-open-letter-2017/ (Accessed 6 December 2021).

Future of Life Institute. (2017b). Asilomar AI principles. Available at: https://futureoflife.org/ai-principles/ (Accessed 6 December 2021).

Future of Life Institute. (2018). Open letter to professor Sung-Chul Shin, President of KAIST from some leading AI researchers in 30 different countries — AI and robotics researchers Boycott South Korea tech institute over development of AI weapons technology. Available at: https://futureoflife.org/2018/04/04/ai-and-robotics-researchers-boycott-kaist/ (Accessed 6 December 2021).

Gagnè, J-F. (2019). Global AI talent report 2019. *jfgagne*. Available at: https://jfgagne.com/talent-2019/ (Accessed 30 January 2024).

Gagnè, J-F. (2020). Global AI talent report 2020. *jfgagne*. Available at: https://jfgagne.com/global-ai-talent-report-2020/ (Accessed 30 January 2024).

Ganghof, S. (2003). Promises and pitfalls of veto player analysis. *Swiss Political Sci. Rev.*, 9(2), 1–25.

Google Employees. (2018). Letter: Project Maven. Available at: https://static01. nyt.com/files/2018/technology/googleletter.pdf (Accessed 6 December 2021).

Google Employees Against Dragonfly. (2018). We are google employees. Google must drop dragonfly. *Medium*. Available at: https://medium.com/@googlers againstdragonfly/we-are-google-employees-google-must-drop-dragonfly-4c8a30c5e5eb (Accessed 6 December 2021).

Google Walkout for Real Change. (2019). When we organize, we win: #GoogleWalkout, one year later. *Medium*. Available at: *Medium*. https:// googlewalkout.medium.com/when-we-organize-we-win-googlewalkout-one-year-later-1c24ad6a2c87 (Accessed 6 December 2021).

Haas, P. M. (1992). Introduction: Epistemic communities and international policy coordination. *Int. Organ.*, 46(1), 1–35.

Hajian, S., Bonchi, F., and Castillo, C. (2016). Algorithmic bias: From discrimination discovery to fairness-aware data mining. In *Proceedings of the 22nd ACM SIGKDD International Conference on Knowledge Discovery and Data Mining*, KDD'16, New York, USA: Association for Computing Machinery, pp. 2125–2126. http://doi.org/10.1145/2939672. 2945386.

High-Level Expert Group on Artificial Intelligence (AI HLEG). (2019). The ethics guidelines for trustworthy artificial intelligence. Available at: https://digital-strategy.ec.europa.eu/en/library/ethics-guidelines-trustworthy-ai (Accessed 30 January 2024).

Hirschman, A.O. (1970). *Exit, Voice and Loyalty: Responses to Decline in Firms, Organizations and States* (Illustrated edn.). Cambridge, MA: Harvard University Press.

Hua, S.-S. and Belfield, H. (2021). AI & antitrust: Reconciling tensions between competition law and cooperative AI development. *Yale J. Law Technol.*, 23, 136.

International Committee of the Red Cross. (2021). ICRC position on autonomous weapon systems. Available at: https://www.icrc.org/en/document/icrc-position-autonomous-weapon-systems (Accessed 6 December 2021).

Jahn, D. (2011). The veto player approach in macro-comparative politics: Concepts and measurement. In *Studies in Public Choice* (pp. 43–68). Springer. Available at: https://ideas.repec.org/h/spr/stpchp/978-1-4419-5809-9_3.html (Accessed 6 December 2021).

Kinoshita, R. (2022). Views of AI PhD recipients on resources to build the domestic talent pool. Center for Security and Emerging Technology. Available at: https://cset.georgetown.edu/publication/views-of-ai-phd-recipients-on-resources-to-build-the-domestic-talent-pool/ (Accessed 17 September 2022).

Koene, A. (2017). Algorithmic bias: Addressing growing concerns [Leading Edge]. *IEEE Technol. Soc. Mag.*, 36, 31–32.

Krizhevsky, A., *et al.* (2012). ImageNet classification with deep convolutional neural networks. *Advances in Neural Information Processing Systems*, 25.

Leung, J. (2019). Who will govern artificial intelligence? Learning from the history of strategic politics in emerging technologies. PhD thesis, University of Oxford. Available at: http://purl.org/dc/dcmitype/Text; https://ora.ox.ac.uk/objects/uuid:ea3c7cb8-2464-45f1-a47c-c7b568f27665 (Accessed 11 November 2021).

Maas, M. M. (2019). How viable is international arms control for military artificial intelligence? Three lessons from nuclear weapons. *Contemp. Secur. Policy*, 40(3), 285–311.

Metz, C. (2018). AI researchers are making more than $1 million, even at a nonprofit. *The New York Times*. Available at: https://www.nytimes.com/2018/04/19/technology/artificial-intelligence-salaries-openai.html (Accessed 6 December 2021).

Microsoft Employees. (2018). An open letter to Microsoft: Don't bid on the US military's project JEDI. *Medium*. Available at: https://medium.com/s/story/an-open-letter-to-microsoft-dont-bid-on-the-us-military-s-project-jedi-7279338b7132 (Accessed 6 December, 2021).

Miller, C. and Coldicutt, R. (2019). People, power and technology: The tech workers' view. London: Doteveryone. Available at: https://doteveryone.org.uk/report/workersview/ (Accessed 6 December, 2021).

Moyes, R. (2020). From "Pink Eyed Terminators" to a Clear-Eyed Response? UK Policy on Autonomous Weapons. London: Article36. Available at: https://article36.org/updates/uk-policy-analysis-2020/ (Accessed 2 February 2022).

Nolan, B. and Coulter, M. (2022). Everyone's fighting to hire machine-learning engineers, with salaries as high as $250,000. We spoke to researchers at DeepMind and the AI unicorn tractable on how to break. *Business Insider*. Available at: https://www.businessinsider.com/how-to-land-job-as-machine-learning-engineer-250000-salary-2022-2 (Accessed 17 September 2022).

Nolan, L. (12 November 2018). Jeff Bezos is wrong, tech workers are not bullies. *Financial Times*. https://www.ft.com/content/f4bd1860-e230-11e8-a8a0-99b2e340ffeb.

O'Sullivan, L. (2019). I quit my job to protest my company's work on building killer robots. *American Civil Liberties Union*. Available at: https://www.aclu.org/blog/national-security/targeted-killing/i-quit-my-job-protest-my-companys-work-building-killer (Accessed 6 December 2021).

OECD. (2021). Tricot, R. Venture capital investments in artificial intelligence: Analysing trends in VC in AI companies from 2012 through 2020. *OECD Digital Economy Papers*, 319, OECD Publishing, Paris. Available at: http://www.oecd.org/digital/venture-capital-investments-in-artificial-intelligence-f97beae7-en.htm.

OECD.AI Policy Observatory. (2022). National AI policies & strategies. Available at: https://oecd.ai/en/dashboards (Accessed 1 February 2022).

Ord, T. (2020). *The Precipice: Existential Risk and the Future of Humanity.* London: Bloomsbury Publishing.

Partnership on AI to Benefit People & Society. (2016). Industry Leaders Establish Partnership on AI Best Practices. Available at: https://deepmind.com/ blog/announcements/announcing-partnership-ai-benefit-people-society (Accessed 6 December 2021).

Payne, K. (2018). Artificial intelligence: A revolution in strategic affairs? *Survival*, 60(5), 7–32.

Poulson, J. (2019). Opinion — I used to work for google. I am a conscientious objector. *The New York Times.* Available at: https://www.nytimes.com/ 2019/04/23/opinion/google-privacy-china.html (Accessed 6 December 2021).

Rosert, E. and Sauer, F. (2021). How (not) to stop the killer robots: A comparative analysis of humanitarian disarmament campaign strategies. *Contemp. Secur. Policy*, 42(1), 4–29.

Russell, S., Aguirre, A., Conn, A., and Tegmark, M. (2018). Why you should fear 'Slaughterbots' — A response. *IEEE Spectrum.* https://spectrum.ieee. org/why-you-should-fear-slaughterbots-a-response (Accessed 6 December 2021).

Salesforce Employees. (2018). An open letter to salesforce: Drop your contract with CBP. Available at: https://fightfortheftr.medium.com/an-open-letter-to-salesforce-drop-your-contract-with-cbp-a8260841b627 (Accessed 6 December 2021).

Siegmann, C. and Anderljung, M. (2022). *The Brussels Effect and AI: Will EU Regulation Shape the Global AI Market?* Oxford, UK: Centre for the Governance of AI (GovAI).

Slaughterbots. (2017). Available at: https://www.youtube.com/watch?v=9CO6 M2HsoIA (Accessed 6 December 2021).

Slaughterbots — If Human: Kill(). (2021). Available at: https://www.youtube. com/watch?v=9rDo1QxI260&ab_channel=FutureofLifeInstitute (Accessed 6 December 2021).

Solaiman, I., *et al.* (2019). Release strategies and the social impacts of language models. *arXiv:1908.09203 [cs].* Available at: http://arxiv.org/abs/1908.09203 (Accessed 2 February 2022).

TenCent Research Institute. (2017). *Global AI Talent White Paper.* TenCent Research Institute.

Tiernan, R. (2019). Google DeepMind's Demis Hassabis is one relentlessly curious public face of AI. *ZDNet.* Available at: https://www.zdnet.com/article/ googles-demis-hassabis-is-one-relentlessly-curious-public-face-of-ai/ (Accessed 6 December 2021).

Tiku, N. (2019). Three years of misery inside Google, the happiest company in tech. *Wired.* Available at: https://www.wired.com/story/inside-google-three-years-misery-happiest-company-tech/ (Accessed 6 December 2021).

Tsebelis, G. (1995). Decision making in political systems: Veto players in presidentialism, parliamentarism, multicameralism and multipartyism. *Br. J. Political Sci.*, 25(3), 289–325.

Tsebelis, G. (2002). *Veto Players: How Political Institutions Work.* Princeton, NJ: Princeton University Press.

TUC. (2019). *The Added Value of Trade Unions: New Analyses for the TUC of the Workplace Employment Relations Surveys 2004 and 2011.* London: TUC. Short Report. Available at: https://www.tuc.org.uk/sites/default/files/The%20 added%20value%20of%20trade%20unions%20RS.pdf.

Upadhya, R. K. (2018). Tech workers, platform workers, and workers' inquiry. *Tech Workers Coalition.* Available at: https://medium.com/tech-workers-coalition/tech-workers-platform-workers-and-workers-inquiry-92fbc 6369647 (Accessed 6 December, 2021).

Verbruggen, M. (2019). The role of civilian innovation in the development of lethal autonomous weapon systems. *Global Policy*, 10(3), 338–342.

Vignard, K. (2018). Manifestos and open letters: Back to the future? *Bull. At. Sci.*, Available at: https://thebulletin.org/2018/04/manifestos-and-open-letters-back-to-the-future/ (Accessed 6 December 2021).

Whittaker, M., et al (2018). *AI Now 2018 Report.* New York: AI Now. Available at: https://ainowinstitute.org/publication/ai-now-2018-report-2 (Accessed 30 January 2024).

Whittlestone, J., Nyrup, R., Alexandrova, A., and Cave, S. (2019). The role and limits of principles in AI ethics: Towards a focus on tensions. In *Proceedings of the 2019 AAAI/ACM Conference on AI, Ethics, and Society (AIES '19)*, New York, USA: Association for Computing Machinery, pp. 195–200. Available at: http://doi.org/10.1145/3306618.3314289 (Accessed 6 December, 2021).

Wilkinson, A., Donaghey, J., Dundon, T., and Freeman, R. B. (eds.). (2020). *Handbook of Research on Employee Voice* (2nd edn.). Cheltenham, UK; Northampton, MA, USA: Edward Elgar Publishing.

Zhang, B., et al. (2021). Ethics and governance of artificial intelligence: Evidence from a survey of machine learning researchers. *arXiv:2105.02117 [cs].* Available at: http://arxiv.org/abs/2105.02117 (Accessed 2 February 2022).

Zweben, S. and Bizot, B. (2019). 2018 Taulbee survey: Undergrad enrollment continues upward. Doctoral Degree Production Declines but Doctoral Enrollment Rises. 72.

Zweben, S. and Bizot, B. (2020a). 2019 Taulbee survey: Total undergrad CS enrollment rises again, but with fewer new majors. Doctoral Degree Production Recovers from Last Year's Dip. 61.

Zweben, S. and Bizot, B. (2020b). 2020 Taulbee survey: Bachelor's and doctoral degree production growth continues but new student enrollment shows declines. 67. Computing Research Association.

Zweben, S. and Bizot, B. (2022). 2021 Taulbee survey: CS enrollment grows at all degree levels, with increased gender diversity. 81.

Zwetsloot, R. and Corrigan, J. (2022). *AI Faculty Shortages*. Center for Security and Emerging Technology. Available at: https://cset.georgetown.edu/publication/ai-faculty-shortages/ (Accessed 17 September 2022).

Zwetsloot, R. and Dafoe, A. (2019). Thinking about risks from AI: Accidents, misuse and structure. *Lawfare*. Available at: https://www.lawfareblog.com/thinking-about-risks-ai-accidents-misuse-and-structure (Accessed 6 December 2021).

Chapter 8

From Evaluation to Action: Ethics, Epistemology, and Extreme Technological Risk

Lalitha S. Sundaram, Matthijs M. Maas, and S. J. Beard

8.1 Introduction

The title of this book is *Managing Extreme Technological Risk*. This refers to a substantial project of vital importance for humanity, tackling the multiple risks that, as Catherine Rhodes set out in her introduction to this volume, cross a threshold (either singularly or cumulatively, in combination or cascades) above which humanity's existence is threatened (known as Extreme Technological Risk or ETR). However, it also refers to a more specific and limited research programme, which was hosted by the University of Cambridge's Centre for the Study of Existential Risk from 2015 to 2020 and funded by the Templeton World Charity Foundation. This programme sought to make a meaningful contribution to this larger aim. The chapters in this book all arose, to a greater or lesser extent, out of the work of this more bounded research programme and this book reflects upon a range of its findings and impacts.

We write this chapter as researchers who started working on managing ETR at different, though still early, stages in our academic careers, from a range of disciplines — philosophy, life sciences, and law — and who have been closely involved in the development of CSER and its work on extreme technological risk.[1] Aspects of our experiences are probably familiar to any scholar entering a new field of study; others will be more specific to our context of managing ETR. We encountered challenges and tensions, as well as the exciting opportunity of engaging in an emerging field with many similarly motivated colleagues. Questions arose for us around why the field appeared (or was perceived) to function in particular ways, and what implications that had for our work and its outcomes.

In this chapter, we aim to combine our perspectives on both the broad research endeavour of managing ETR and the smaller CSER programme of Managing Extreme Technological Risk to revisit some of the key questions and points of reflection that arose during that work, how these were addressed (both during and after the programme), and what questions we feel remain unanswered and in need of further consideration. We hope that this will offer some valuable insights both to those who are already engaged with tackling the challenges posed by ETR who would like to take a more reflexive approach to their work and to anyone new to this field who wants to understand how it is developing and which areas could benefit most from further engagement and novel perspectives.

8.2 A Brief History of Managing Extreme Technological Risk at CSER and Beyond

The past two decades have seen rapid growth in the magnitude and variety of technological risks facing humanity (Deudney, 2018), from the onset of fossil-fuel-driven climate change (Beard *et al.*, 2021) to a new era of nuclear instability (Krepon, 2021) — from advanced biotechnologies (Noun and Chyba, 2008; Kemp *et al.*, 2020) to the emergence of widespread, increasingly capable, and potentially misaligned artificial

---

[1]The authors of this final chapter have all been researchers at the Centre for the Study of Existential Risk but have different relations to the Programme on Managing Extreme Technological Risk. SJ Beard was one of the core researchers on this, focusing on the evaluation of extreme technological risk. Lalitha Sundaram was a close collaborator whose primary role was external but who was able to provide important insights to many of its researchers by offering more of an outsider perspective. Finally, Matthijs Maas provided later engagement with researchers in the programme, having joined the centre after METR had been completed, but having been active within the ETR community in the intermediate years. Nevertheless, there were many important points of dialogue between his own research and experience and that of METR programme researchers.

intelligence (AI) systems (Ord, 2020; Christian, 2020). This has been accompanied by growing awareness of and concern about these threats among the public and policymakers at national and international levels (Global Challenges Foundation/Novus, 2020; House of Lords, 2021; United Nations, 2021).

The initial idea behind the Managing Extreme Technological Risk research programme was that in order to meet this rising challenge it was necessary to undertake 'academic engineering': 'to design and establish a new interdisciplinary subfield, engineered for the specific purpose of managing ETR in the long term'.[2] This idea, and the programme that emerged from it, initiated an important evolution in the way that ETR is studied and understood that formed part of a wider 'second wave' of ETR research (Beard and Torres, 2020). Up to this point, the study of ETR had tended to be split. Some teams focused on the mitigation of specific risks (such as nuclear war). Other cross-cutting risk work emerged from a few small research centres with, at that point, the specific aim of helping humanity safely navigate towards what they saw as a state of 'technological maturity' where technologies, such as AI, space colonisation, human enhancement, and nanotechnology, would be able to eradicate external threats and provide for unparalleled levels of abundance and well-being (Bostrom, 2013). As was noted by the METR project description, however, at the time, none of these efforts constituted a 'systematic attempt to answer the question as to how this class of risks can best be identified, managed and mitigated, in the long term'.[3]

The METR programme sought to respond to and remedy this gap. While the academic engineering approach was specific to CSER, this desire to innovate new holistic approaches to managing ETR was also shared by a number of other efforts across the field, including existing ETR research organisations, like the Future of Humanity Institute and Machine Intelligence Research Institute, new centres looking to undertake work in this area, such as the Future of Life Institute and Global Catastrophic Risk Institute, and others who wished to contribute to the larger endeavour of managing ETR, many of whom were aligned with the burgeoning Effective Altruist community.[4]

---

[2] This, and some subsequent, quotes are taken directly from the Managing Extreme Technological Risk project description, a copy of which can be provided on request.

[3] Managing Extreme Technological Risk project description.

[4] At least part of this concern can be attributed to the influential (Bostrom, 2013).

The way in which the Managing Extreme Technological Risk programme aimed to make progress in this area was by bringing together an interdisciplinary community of scholars who could bring specialist contributions to the field, while retaining strong connections to their home disciplines. These researchers would seek to build an interdisciplinary community that 'sits at the centre of a web with "horizontal" links into a wide range of other disciplines' while also building 'practical links both "downwards" into technology development communities, and "upwards" into the policy and public arena'.[5] This is illustrated via Figure 1 (affectionately known within the programme as the 'starfish').

The initial community consisted of five core researchers working with a variety of partners, advisors, and collaborators. As they developed their research strands over the course of the METR programme and engaged with the wider work of the Centre, several questions and challenges arose about the nature of the interdisciplinary 'science' of ETR that the programme aimed to establish (and arguably achieved). Many of these originated from the researchers' own disciplinary backgrounds in fields such as responsible innovation and the philosophy of science. For instance, at an away day in July 2017, the group undertook a red team/blue team exercise developing and responding creatively to potential critiques of the programme's aims and methods.[6] Key topics at this exercise included the following:

Challenges of understanding: Whether one can credibly make predictions about extreme events, whether risk generalists can add anything to fields where technical specialists already exist, and how to avoid risks of falling prey to pseudoscience.

Challenges of scope: Whether we should focus on more likely and/or immediate threats and whether there are important risk areas we are not yet aware of.

Challenges of impact: Whether we can find the right people to influence, whether our work can get noticed, whether we can work against existing incentive structures, and how we might handle our relationships with powerful actors.

---

[5] Managing Extreme Technological Risk project description.
[6] The group included two of this chapter's authors, SJ Beard and Lalitha Sundaram; this chapter builds from their insights from this exercise.

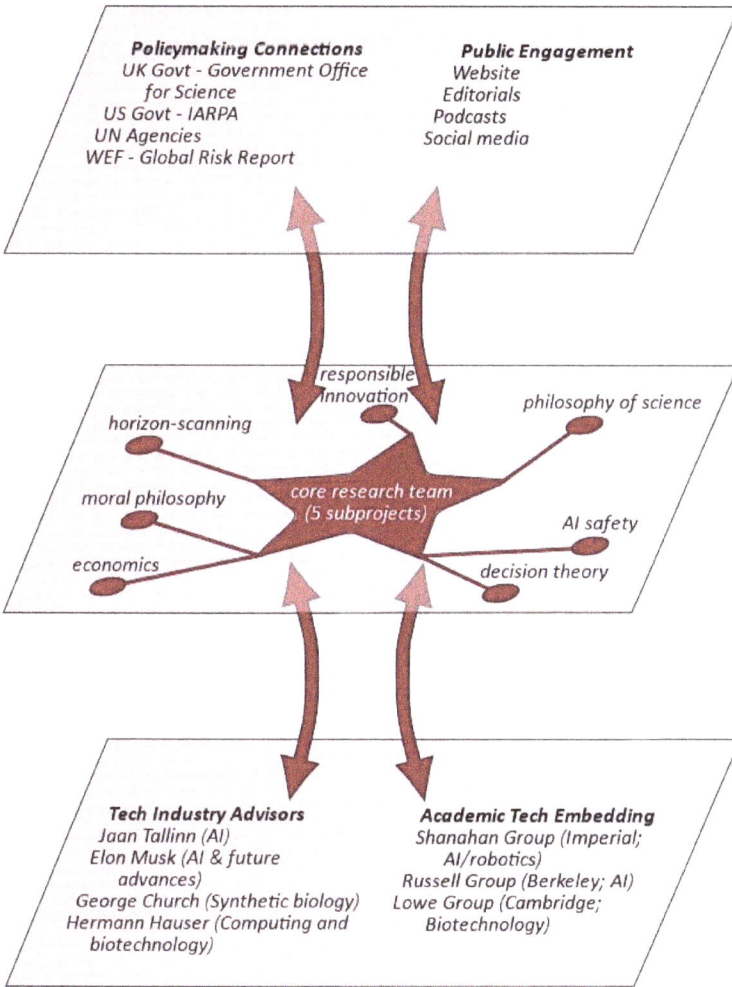

Policymaking Connections
UK Govt - Government Office
for Science
US Govt - IARPA
UN Agencies
WEF - Global Risk Report

Public Engagement
Website
Editorials
Podcasts
Social media

responsible
Innovation

philosophy of science

horizon-scanning

moral philosophy

core research team
(5 subprojects)

AI safety

economics

decision theory

Tech Industry Advisors
Jaan Tallinn (AI)
Elon Musk (AI & future
advances)
George Church (Synthetic biology)
Hermann Hauser (Computing and
biotechnology)

Academic Tech Embedding
Shanahan Group (Imperial;
AI/robotics)
Russell Group (Berkeley; AI)
Lowe Group (Cambridge;
Biotechnology)

Figure 1. The Starfish Model.

This exercise and other reflective discussions led to various activities in response to these concerns, including a review of the existing evidence base in existential risk (Beard *et al.*, 2020), wide-ranging collaborative projects with diverse partners, such as on the malicious use of AI (Brundage *et al.*, 2018), and trustworthiness in AI development (Brundage *et al.*, 2020), and a range of impact and engagement projects such as establishing an All-Party Parliamentary Group for Future

Generations in the UK. These activities highlighted the value of such reflective discussions, especially with regard to CSER's core values and purpose. As well as identifying key challenges, the red team/blue team exercise and other reflections also identified some of our chief assets for responding to them. These included the following:

Our clear mission statement to mitigate risks of human extinction and civilisational collapse: These will consistently drive our decisions, are easy to understand and explain to others, overlap with a wide range of other concerns, and align with the interests of many powerful actors.

Our existing and potential future network: This provides us with access to expertise that we lack, oversight on our decisions and priorities, and a natural feedback mechanism should anything go wrong.

Our diversity as a group, in terms of knowledge and expertise, identity, background, and networks.[7]

CSER took the opportunity to further explore these challenges and opportunities with the broader ETR community at the second Cambridge Conference on Catastrophic Risk, held by CSER in April 2018. Importantly, the wider community was invited to attend and engage in reflection on these issues with us; this was more than an internal exercise. The Conference focused on challenges of evaluation and impact; challenges of evidence for existential risk research; challenges of scope and focus (what might we be missing/neglecting?); and challenges in communication and responsible engagement. In considering these, the conference highlighted many shared issues, with five key themes emerging in need of further work:

> How can we best think about issues that are hard to think about, such as 'black swans'[8] and 'unsexy risks',[9] and how can we persuade other people to do the same?

---

[7] Quote taken from an internal report of the event prepared by SJ Beard, Lalitha Sundaram, and Shahar Avin.
[8] See Peter (2017).
[9] See Kuhlemann (2018).

Given the broad interdisciplinary nature of our work, how can we help individual researchers understand how their work contributes to managing ETR as a whole and help researchers collaborate towards achieving this common aim?

How can we make our research truly relevant, both to the scale of the problems that we face in managing ETR and for everyone who has a stake in creating and/or mitigating them?

How can we take advantage of the diversity of knowledge, perspectives, and opportunities available to us while avoiding conflict and retaining our ability to make collective decisions?

How can we deal with the unintended, and sometimes perverse, consequences of studying ETR, which might include sensationalist press coverage and unintended policy impacts or the locking-in of faulty assumptions and bad decision-making structures to our future understanding of risk?

In responding to these challenges and questions, conference participants tended to emphasise the need to give greater consideration to what reaches the agenda of the research community and why; pay attention to how the community can foster diversity (in the disciplines it includes, the approaches it adopts, and the people with whom it engages); enforce some form of boundary setting that retains the community's coherence and focus; choose what and how the community will measure, evaluate, and collate evidence when assessing risks and their management; and understand where changes might need to be made and identify practical actions to take, who one needs to influence, and how.[10]

These insights influenced the Managing Extreme Technological Risk programme and helped to inform CSER's thinking about where further work was needed beyond it. Among other work, this informed the creation of a successor programme to develop A Science of Global Risk (ASOGR), also funded by the Templeton World Charity Foundation. A Science of Global Risk shared the founding vision of METR, further developing it by focusing on what's required — particularly in terms of communication and outreach — to bring together and sustain a community devoted to managing extreme risks, and to strengthen the connection between understanding and action alongside key stakeholders in policy, industry, and the public.

---

[10] A fuller discussion of how participants addressed these challenges and themes can be found in our conference report, Beard and Rhodes (2018).

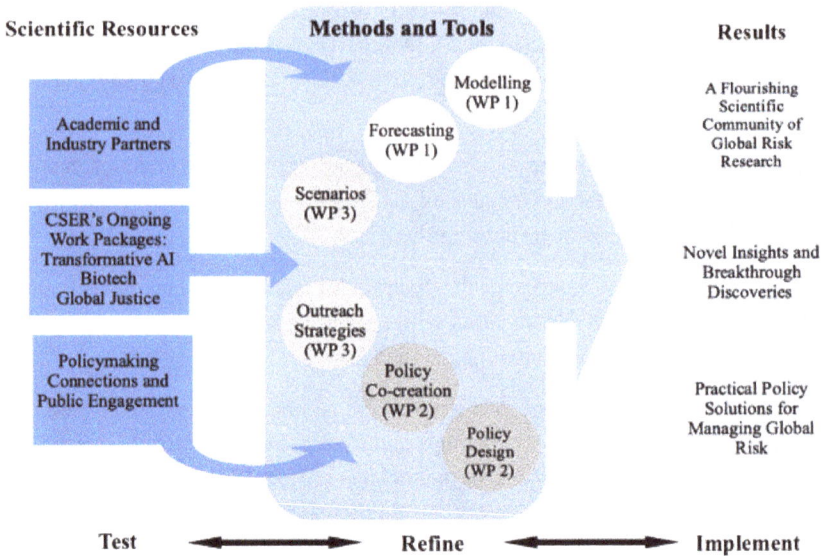

Figure 2.   An Integrated Science of Global Risk.

The result was a research programme built around continuing to develop CSER's approach to foresight and horizon scanning as tools for interdisciplinary engagement with ETR (building on the work of METR's subproject 2) alongside work on developing new approaches for communications and engagement, and refining CSER's theory of change and using policy co-creation and design techniques to build new and more collaborative connections with policymakers to mitigate ETR. This new programme is illustrated in Figure 2.

Final edits to this chapter were made after the ASOGR programme had successfully completed, and there are no plans for a direct successor project to replace it on a like-for-like basis. The Centre currently contains researchers with a wide variety of interests and disciplines and maintains collaborations with a still wider range of partners, at other existential risk research organisations, academic departments, communities of practice, and anywhere else people work to make humanity safer. Recent examples of this have included establishing a science policy interface with the Simon Institute for Longterm Governance, which brings together scientists and policymakers from around the world to discuss reducing ETR; an intensive visitor programme, which brought together ten diverse visitors to the Centre for ten days to discuss the science of global risk and its future; and Intelligence Rising, a tabletop scenario role-playing exercise that allows any interested party to participate in and think seriously about the future trajectories and risks from artificial intelligence in an interactive way. Some of this, and related work, has been collected in an introduction to the science of global risk and an anthology of recent work by the Centre, co-edited by one of this chapter's authors (Maas, 2022). As such, CSER might now be seen as forming part of a more amorphous community of alignment, where diverse researchers and groups collaborate in many different ways to align their research with other stakeholders (including researchers, developers, policymakers, and activists) who share the same aim of making the greatest contribution to our understanding and mitigation of ETR.[11] This is, of course, entirely in keeping with the original method of academic engineering.

Nevertheless, many of the questions, themes, and challenges that have emerged during the METR programme, its successor, and beyond are yet unresolved and remain important for this wider community to consider. At present, the community appears to be transitioning to a 'third wave' of research, where the increased diversity of perspectives from the second wave has led to the creation of new methods and approaches for the study and management of these risks. This is manifest in both the emergence of a transdisciplinary field for studying ETR, often known as Existential Risk Studies, and an increasingly diverse range of interdisciplinary collaborations with people from other disciplines and backgrounds. In turn, this has prompted further growth and diversification, and has led to many established assumptions in the field being questioned and alternatives developed. It is thus highly likely that even more questions will arise in future. In what follows, we present seven questions that we feel anyone who identifies as part of this community (or would like to) should reflect on.[12]

In the same way that ETR researchers, within CSER and elsewhere, have often urged others to innovate responsibly in science and technology, our view is that such reflection, discussion, and possible revision of approaches and activities should be an ongoing, dynamic, and iterative process for any developing field of study. As such, we do not seek to

---

[11] Though of course there is still great diversity in who contributes, and what those aims are, as we discuss below in Question 3.

[12] The authors explored some of these questions, together with others, as a CSER panel in July 2021 (CSER, 2021). A longer discussion of these questions is also available (Sundaram *et al.*, 2022).

provide definitive answers to these questions, but instead examine our experiences, and other views we've encountered, in order to prompt further discussion and exploration. There are good reasons why others might approach and answer these questions in different ways, and there might be strong disagreements in some areas.

8.3 Seven Questions about Managing Extreme Technological Risk

The questions we explore in this section arose from individual reflections and group discussions within CSER, and with others in the ETR community at events such as our conferences and workshops. While different in focus, they point to common challenges around research direction, pluralism, epistemology, diversity, inclusion, representation, and accountability within the field. As such, we believe they represent 'crucial considerations' (Bostrom, 2014a) in charting a path forward. They connect to the discussion of such issues in other chapters and point to important issues for the community to reflect on.

8.3.1 Q1. How do we choose which extreme technological risk to study?

At a straightforward level, our choices as individual researchers about which research topics and cases to address were the result of an iterative combination of various factors. Like other researchers, we are influenced in such choices by internal and external considerations. Internal influences include our personal convictions about the world's problems and how they motivate our work, as well as the existing disciplinary expertise we bring to managing ETR, and our views on how we can most effectively contribute from that basis. External influences include those related to academic career structures, such as how certain lines of research are perceived by others, the professional academic pressure to publish in particular outlets and be able to demonstrate impact for career advancement, and the ability to attract funding. Like anyone else, researchers are also subject to a range of other influences on decision-making, such as cognitive biases and perceptions of social capital and esteem.

Personally, our choices were also shaped by the design of the METR programme, and the initial guiding questions outlined for its research strands. However, given the expanding range of ETRs facing humanity that need studying, we were faced with further questions about how to prioritise and select among them. As our work progressed, we were able to identify particularly productive case studies as topics gained wider public or policy attention, as we were able to engage specific expertise, and as we were to join collaborative efforts with others. These included geoengineering as a cross-cutting case study (Currie, 2018; Halstead, 2018); engaging dual-use research of concern as it reached policy agents (Rhodes, 2020; Husbands, 2018); joint work on 'Malicious uses of AI' (Brundage *et al.*, 2018); and on military AI — synthesising expertise on AI risk, military technologies, and the nuclear realm (Maas *et al.*, 2023; Avin and Amadae, 2019). Likewise, having realised that initiating horizon-scanning work with an attempt to capture all ETR was impractical, we chose the area of bioengineering as a starting point, as we had good links to relevant technology and policy communities, as well as this being a clear area of connection to high-level policy interest in its outputs (Wintle *et al.*, 2017; Kemp *et al.*, 2021). The selection of case studies had a practical basis: an intersection of interest, need, opportunity and our comparative advantage, as well as remaining guided by the original selection of research areas for the programme.

An early piece of collaborative work for the METR programme and Centre highlighted different categories of risk the world might currently face, with critical system failures, global spread mechanisms, and mitigation fragilities combining to imperil humanity (Avin *et al.*, 2018). This means that questions of prioritisation featured not only between particular sources of risk, or their impacts in particular domains, or potential for intervention by particular communities, but also — connecting with wider discussions in the ETR community about the importance of considering vulnerabilities and exposures to risk — with whether and how to prioritise new ways of studying and mitigating ETR in general, rather than addressing individual ETR in separation (Liu *et al.*, 2018).

Utilising some form of prioritisation is necessary for individual researchers, research institutions, and the ETR community as a whole, as limited resources need to be targeted. However, in areas subject to significant uncertainty, prioritisation efforts carry risks and present major challenges. In Chapter 2, Sundaram and Currie outline some of the benefits and risks associated with methods of prioritisation for a science of ETR, and question the extent to which systematic prioritisation might ultimately be unworkable in this context. Nevertheless, the authors believe that continued reflection by the community on how we approach selection and prioritisation has value in and of itself, and that acknowledgement of the challenges such as those Sundaram and Currie present should encourage some diversity in approaches to prioritisation.

In the wider community of those working on managing ETR, there is widespread use of explicit 'cause prioritisation' frameworks which use certain heuristics or principles to guide decision-making. The 'Importance, Tractability, and Neglectedness' (ITN) framework is a particularly influential example. This evaluates research areas not only according to their importance or significance (considered to be a product of the number of people affected by a risk, the magnitude of its impact, and its certainty or probability) but also according to the tractability of mitigating risks in this area, and the neglectedness of the problem in terms of the resources and attention that are allocated to it today (or are likely to be allocated to it in future, before it is too late). This can be captured in a set of simple questions or heuristics. As Winter *et al.* put it in their discussion of the framework, a problem's 'importance' can be understood as follows: 'If we solved this problem, by how much would the world become a better place?' Tractability refers to the possibility of actually solving the problem. As a heuristic, one might ask: 'If we doubled direct effort on the problem, what fraction of the remaining problem would we expect to solve?' Finally, neglectedness refers to the question: 'How many resources will be dedicated to solving the problem before it is too late?' (Winter *et al.*, 2021). As such, under the ITN framework, while any existential risk is *prima facie* important, the size of impact alone is not enough to warrant prioritisation, if the issue already receives considerable attention and/or if making any meaningful progress appears intractable (relative to other existential risks, or relative to any other societal problems generally).

In a world with limited resources and many uncertain challenges to be addressed in parallel, there is an underappreciated value to the ITN framework. However, while attractive and innovative, there persist a number of potential problems with its application. Firstly, some ITN terms, such as a problem's tractability, may be very hard to assess *ex ante*. While there are situations where we could base ballpark estimates for the full costs of successful ETR mitigation interventions (i.e., a component of considering its 'tractability') on prior information, there may be low-information environment ETR (such as those concerning unprecedented technologies or complex political systems) where it is not until a problem has been solved that we can actually know what amount of effort or resources it took to solve it. Secondly, assessments of tractability and neglectedness are very sensitive to how a problem is framed or positioned in specific fields or contexts. For instance, some ETR scholars have de-emphasised work on climate change, not because they view it as an unimportant risk, but because they view it as a relatively non-neglected issue, given the size of the climate science field as a whole and its global public attention (e.g., Todd, 2017). Within climate change research, however, the potential for climate change to be an existential or global catastrophic risk is in fact very neglected, with the scientific focus squarely on higher-probability but lower-impact climate trajectories (Beard *et al.*, 2021; Jehn *et al.*, 2021). A third problem for the ITN framework is the influence of cognitive biases on assessments of all three terms. Given deep epistemic uncertainty, and given different moral intuitions as well as differences in scope (in-)sensitivity around ETR, there may be a lot of room for individual and group biases to affect which problems are seen as priorities and thus to be perpetuated within research without sufficient challenge.

Kuhlemann (2019) provides an alternative perspective on how people prioritise between different types of ETR, arguing that certain biases have led the ETR community to be overly concerned about problems that share certain features: epistemic neatness (making it easy to identify which disciplines are best suited for studying them); a sudden onset (in that they 'crystallise abruptly, with obviously catastrophic outcomes from as little as a few hours to, at most, a few short years'); and being technologically driven (having 'a close relationship with rather flattering ideas about human ingenuity and intellectual prowess'). Kuhlemann characterises risks with these three properties as 'sexy' and contrasts them to the many 'unsexy' problems that are complex, slow moving, and relate to more fundamental issues with human society organisation — such as issues of overpopulation and environmental breakdown — and which are thus less 'interesting', even if they may be no less important. On the other hand, it should be noted that in focusing on overpopulation as the single biggest cause for a range of 'unsexy' risks, Kuhlemann herself seems to focus on a neat and simple problem that is easily attributable (to the post-second world war 'population bomb' that has seen the human population quadruple in a century) and that focuses on individual reproductive choices (both of which suggest a relatively high degree of human agency and have their own 'sexiness'). The case of paying more attention to issues of 'creeping complexity' is well taken but, in developing cause prioritisation mechanisms for considering complex and systemic risks, we nevertheless need to be careful to not fall into the same reductionist modes of thinking that we found problematic in other ways of viewing the world.

The specific ETR examples that the METR programme clustered around in its work — advanced biotechnologies, AI, geoengineering, nuclear security, and to a lesser extent food security, climate change, and ecological collapse — are similar to the portfolio of the wider ETR community. This is not necessarily problematic, but it does highlight the need to consider how and why we are focusing on particular topics and whether this might indicate neglect of other key areas.

8.3.2 *Q2. How should or could ETR mitigation be weighed against addressing other global priorities?*

One of the earliest concerns of researchers in the Managing Extreme Technological Risk programme was how to balance the study and management of ETR against other pressing challenges. All of the original researchers in this programme came from long-established disciplines with their own, sometimes quite different, priorities (for instance, addressing the pressing global health needs rather than preventing the worst possible public health disasters), and it was readily apparent that these might sometimes be in conflict with managing ETR. As our work has progressed, and the community focusing on ETR has grown, such concerns have tended to diminish, since more and more of our researchers now see themselves as community 'insiders', who are normalised to ETR research. Yet, they certainly have not disappeared. Indeed, one key consideration in favour of prioritising ETR is that, even though the field has grown in recent years, it still remains comparatively small and seeks to continue pulling in researchers and expertise from other fields.

Another important consideration is explored by Huw Price, in Chapter 3, where he argues for what he calls the 'Douglas Doctrine', according to which we need to consider not only the likelihood of an error in imagination or decision-making occurring but also its impacts. As Price quotes, Douglas's argument is that:

> Although the error rates may be the same in two contexts, if the consequences of error are serious in one case and trivial in the other, we expect decisions to be different. Thus the emergency room avoids as

much as possible any false negatives with respect to potential heart attack victims, accepting a very high rate of false positives in the process. ... In contrast, the justice system attempts to avoid false positives, accepting some rate of false negatives in the process (Price, this volume).

In the case of ETR, as Price argues, we seem to have a paradigmatic case in which we should accept a very large number of false positives if that is what it takes to avert a global catastrophe. On the other hand, we might also question whether there is any real tension between working on managing ETR and other pressing global challenges. As clarified in our earlier definition, ETRs are not only future technological risks but also extend to risks from long-established and deployed technological infrastructures, such as nuclear weapons or fossil fuel technology. Moreover, scholars of ETR have in recent years argued a 'common sense case' (Wiblin and Harris, 2021) for mitigating existential risks on the basis that, far from being extremely unlikely and far away, many types of risk may in fact have small but surprisingly substantive annual probabilities, accumulating to distressingly high probabilities over our lifetimes, and accordingly their mitigation would pass a simple cost–benefit analysis. Other researchers have challenged the binary between 'near-term' and 'long-term' problems in a more fundamental way.

From algorithmic surveillance to the climate crisis, it is often the case that, at least to begin with, the negative impacts of technology fall earliest and most disproportionately on those who are worst off and most marginalised. While presently occurring at scales far lower than those of an existential or global catastrophic risk, such costs can be seen as the canaries in the coalmine, highlighting how current deficits in governance presage likely failures in managing and mitigating eventual more extreme risks, or foreshadowing how rapidly developing technologies are already very poorly aligned with human values, which could give us reason to suspect that their further development may not radically or suddenly dissolve this problem, but instead scale up their harms even further, barring urgent action to curb or re-direct these trends today. Finally, there is the possibility of co-benefits: interventions that can help us address existing challenges can often help us prevent, mitigate, or recover from future catastrophes, and vice versa. Maintaining robust and resilient global systems, redressing injustice and inequality, reinforcing norms around shared values, improving governmental risk and resilience strategies, and increasing international cooperation are all goals that can contribute to a better world today, but they also reduce the likelihood of severe harm from ETR down the line. Poverty, ill health, and ignorance are not only major sources of global misery but also limit our potential as a species to collectively solve our problems. In a sense, humanity is currently in the position of trying to save itself with one hand tied behind its back. Moreover, how we deal with non-existential but still catastrophic threats is likely to provide lessons (or warnings) about how we will deal with existential ones, the COVID-19 pandemic being an example (Liu *et al.*, 2020; Rietveld *et al.*, 2022).

8.3.3   *Q3. What are our approaches to studying and managing ETR?*

In general, the METR programme was pluralistic in its approach to managing ETR, with each of the researchers bringing very different perspectives from their home disciplines (law, ethics, philosophy of science, conservation, and decision theory), often leading to creative discussions and disagreements between them. CSER generally did not have a set approach for studying ETR but sought to learn from each of researchers and to test, refine, and implement innovations that arose from different forms of interdisciplinary collaboration between them.[13] We feel that this experience can serve as a useful example for the wider field.

Within the ETR community, approaches to studying ETR are often thought to be rooted in a set of common assumptions including consequentialist morality and Bayesian epistemology. However, such assumptions are certainly not necessary to engaging in managing ETR. For instance, in Chapter 5, Beard and Kaczmarek (this volume) challenge at least one of these assumptions by developing a contractualist approach to the evaluation of ETR, while in other work, Beard *et al.* (2020) survey the many different methodological approaches to studying ETR.

Nevertheless, there are several overarching approaches to ETR that have been developed in recent years and that serve as the basis for new forms of transdisciplinary research. At present, one can distinguish between at least four different approaches that seem to be playing a significant role in shaping the field. These can be distinguished according to

---

[13] A key early instance of this being Avin *et al.* (2018).

their views on (1) the role of ethics in our understanding of ETR and (2) the most appropriate theory of change for managing such risk. On the former question, around the role of ethics, we distinguish between paradigms that (implicitly or explicitly) see managing ETR as tied up with specific ethical worldviews and those that aim instead to decouple the study of ETR from any particular ethical programme. On the latter question of the most appropriate 'theory of change', we distinguish between views that emphasise 'inside working' (working as a trusted partner with established institutions or frameworks or actors, including those who may be directly or indirectly implicated in the creation of risks, such as with militaries operating nuclear arsenals) and those that focus on 'outside working' (making independent statements about the nature of risk without concern about how these might be received by such key actors or how they might affect prospects for direct influence).

The oldest, and perhaps best known, paradigm for studying ETR is that of the 'concerned scientist'. Ever since scientists involved in the Manhattan Project initiated projects like the Bulletin of the Atomic Scientists (Beard and Bronson, 2023), there has been a strong tradition of concerned scientists and researchers seeking to prevent technological catastrophes by spreading knowledge about the risks inherent in their own field and calling on politicians to act to promote more (technological) safety. Often, members taking this perspective do not start with a concern about existential risk per se or in the abstract, but rather have developed a deep understanding of some scientific issue or domain, which leads to them being concerned about a threat related to that field in particular. Only further along may they potentially find common cause with others in the ETR community. As a result, work in the 'concerned scientist' paradigm is more diverse than those that follow, yet often shares a deep commitment to pluralism and democracy, and a theory of change that focuses on marrying an openness to inside work and a willingness to move to outside work if inside channels are perceived as ineffective. Historically, much of this work focused on the risks of nuclear war, though there have always been those concerned about other technologies as well, from the environmental impacts of industrial farming (Carson, 1962) to the potential for runaway AI (Good, 1964) or nanotechnology (Drexler, 1986).

Subsequent paradigms, that we outline in the following, were developed or adapted specifically for the study and management of ETR as a class. The first, 'longtermist' approach to managing ETR has been a prevalent and particularly influential view within the study of ETR.

Inspired by the philosophy of Derek Parfit (McLaughlin, 2022), it developed through the work of Nick Bostrom, Eliezer Yudkowsky, and other early contributors in the Rationalist community (Chivers, 2019; Moynihan, 2020). This included early work that first coined terminology around existential risks and linked it closely to technological development (Bostrom, 2002). The philosophy also influenced notable key works on existential risks, such as Ord's *The Precipice* (2020), and it has found its most definitive statement in the recent *What We Owe The Future* (Macaskill, 2022b).

The longtermist view sees managing ETR as closely aligned to, and entwined with, particular ethical values or axiologies. Some of this work is motivated by a 'strong longtermist' form of consequentialism (or 'time-neutral' total aggregate utilitarianism) like basic ethical theory. While an important school of thought, the ethical assumptions in the longtermist school of ETR studies are not exhaustive of all ethical perspectives on existential risks; indeed, they are also likely (for better or worse) not representative of global public ethical views on these questions (Schubert *et al.*, 2019; Cremer and Kemp, 2021) nor, for that matter, of moral philosophers. For instance, surveys of the Effective Altruism (EA) community (which is not entirely coterminous with the ETR community, but which has advanced significant concern for existential risks) suggest that its members put greater stock on consequentialist theories of ethics, more than do moral philosophers (Cremer, 2020). Moreover, in the longtermist approach, both the threat model of ETR and the ethical grounds for why these risks should be mitigated are closely connected to the anticipated risks and promises of emerging technologies. In particular, technologies such as AI, robotics, and nanotechnology are seen as centrally important to the future of humanity, for both good and bad (Karnofsky, 2021a).

As such, the longtermist approach to ETR on the one hand still focuses a lot on attempts to 'steer' the direction and pace of technological progress (Karnofsky, 2021b) — in particular paying attention to ways in which defensive or protective technological developments might be sped up, and offensive or dangerous technologies slowed down, in order to avoid risky combinations. However, in contrast to its strong links in terms of ethics, the longtermist approach to ETR has not yet articulated a conclusively determined theory of change. Broadly speaking, it is relatively neutral between 'inside working' as a trusted partner with established institutions and 'outside working' where most of its researchers focus on purely academic research, but may have a strong desire to influence groups who are open to collaboration, such as major governments, tech companies, or other key stakeholders.

A second approach that has emerged more recently tries to decouple the study of ETR from any particular ethical program, studying phenomena like human extinction and global catastrophic risk from an analytical point of view, and leaving the normative implications or recommendations to separate analysis (sometimes called Existential Ethics). Researchers in this paradigm tend to be interested in technological risk from a systems theory approach (Avin *et al.*, 2018; Kreienkamp and Pegram, 2020), perceiving risks as emerging from the interaction between technological developments and critical systems required for human survival and flourishing (Liu *et al.*, 2018; Manheim, 2020). Some who share this approach are also committed to a theory of change based on outside working, which they claim helps to preserve academic integrity and independence from the influence of third parties with a vested interest in these systems, while taking a critical stance towards those they see as primarily responsible for producing such risks. These researchers may point to the extensive history of corporate action aimed at downplaying or obfuscating risks, and the resulting important role of independent research in overcoming this (e.g., Oreskes and Conway, 2011). Others, however, take a more neutral or pragmatic view to combining different theories of change, or see inside working as important to actively correct for other biases (such as ideological commitments or an oversimplified theory of institutional behaviour).

A third approach sees ETR as primarily a matter of security. This paradigm draws close connections between ETR and longstanding work in already securitised fields such as biosecurity, information security, sustainable security, or nuclear security. Although there are many different approaches to the securitisation of existential risk, one particularly prominent recent example is the development of the concept of 'existential security' (Sears, 2020). This combines a concern about extreme technological risk with existing security paradigms and a focus on removing specific vulnerabilities and threats. It may involve scholars aiming to work within and with national security actors. The benefit of a securitisation viewpoint is a greater leeway to engage with existing security-based institutions in governments and militaries. These actors often already work on many aspects related to extreme technological risk; they may be the critical or even the only gatekeepers in achieving any mitigation in some risk areas, or they may at least be able to bring to bear unusual resources, expertise, or political attention. In some politicised issue areas, the involvement of national security actors could even provide political capital or credibility that stretches across prevailing party lines or political coalitions (Klare and Perry, 2021). However, in engaging with military institutions, the paradigm opens itself up to many pre-existing points of criticism about such actors, such as their tendency to be overly threat focused, their dependence on the military–industrial complex, or the risk that security frame solutions might impart zero-sum thinking and unilateral national policies that undercut global cooperation on key risks.

While we see that there is already quite a high degree of diversity amongst ETR scholars about many questions concerning the study and management of ETR, that diversity could likely still be substantially increased were a greater plurality of voices, perspectives, and institutions to enter the field. Indeed, while it may often appear appealing for an ETR community to work with those that share or agree on a series of fundamental assumptions or reference points, this risks breeding implicit assumptions or blind spots. We thus believe that there is no one best approach to ETR and that the traditions of both independence and collaboration that have emerged in the field are worth continuing, even if their different theories of change can sometimes create tensions between researchers and their intermediate objectives. Ultimately, the four approaches we have sketched might once have been somewhat foundational, but should not be considered archetypical. As the disciplines involved have expanded, and the number of people involved has increased, we should expect individual researchers to draw from these (and other) traditions, along with bringing in new approaches to the area. In general, a likely direction of travel for managing ETR will be the continued increase in approaches and underlying motivations or ethical frameworks. In doing so, the field should critically consider the question of what this will mean: how (or even whether) to ensure a level of coherence and consistency in what the field is — and whether it even should, or can, be sustained as a 'field'.

8.3.4 Q4. How do we address the desire to be scientific within ETR research?

As Sundaram and Currie describe in Chapter 2 of this volume, there are many tensions inherent in the desire for those working on managing ETR to be scientific. The field is both devoid of an uncontested evidence base and intimately tied up with questions about policy and impact. It is thus necessary for the field to be highly creative in its approach and to find good ways to explore and evaluate the subjective opinions and good judgement of individual researchers while still being as transparent and objective as possible in its working.

One approach that was developed as part of the Managing Extreme Technological Risk programme was to make our research more methodologically rigorous with the use of futures, foresight, and horizon scanning, terms that describe a wide variety of techniques. The fundamental aim in each case is to provide a systematic way of thinking about future events that allows the judgements of multiple people to be compared and combined. These approaches aim to move beyond individual speculation about what might happen to describe a possibility space or consensus judgement that draws together all of the available information while filtering out individual biases and other sources of 'noise'. However, given the extreme nature of the risks that we work on, CSER has had to adapt these tools to our purposes. As Luke Kemp argues in Chapter 4 '[m]any forecasting tools have proved effective and dependable in making good probabilistic judgements about near-term, well-bounded events', but these are not suitable for managing ETR, which is long term, complex, and ambiguous. Thus, the METR programme and its successor instead 'concentrated on understanding and prioritising plausible developments' by 'using horizon scanning, particularly through the IDEA protocol, to identify emerging issues [in technologies with the potential to create ETR]'. CSER has used this technique to explore emerging issues in biological engineering (Wintle *et al.*, 2017; Kemp *et al.*, 2020) and in UK biosecurity (Kemp *et al.*, 2021).

Another popular technique for understanding extreme technological risk is Superforecasting, developed by the Good Judgment Project (Tetlock and Gardner, 2016), whereby large groups of interested but not necessarily specialist participants are trained in tools such as overcoming cognitive bias and forecasting best practice and then tasked with answering questions about geopolitical events. These responses are algorithmically aggregated and, in some competitive exercises, this method of forecasting consistently prevails even over the judgement of domain experts or professional intelligence analysts with access to classified information (Tetlock *et al.*, 2017).

In a further development of these techniques, one perhaps less interested in 'absolute accuracy of prediction' but rather in exploring the possibility space, CSER has also begun to develop interactive scenario-based tools and role-playing exercises which serve to help understand different stakeholders' knowledge of and relationship to catastrophic risks, in order to encourage participatory technology assessment engagement from a wider group of experts as well as citizens (Avin *et al.*, 2020; Cremer and Whittlestone, 2021; Davies, 2022).

Other scientific techniques also lend themselves to the management of ETR and have been applied within the field with greater and lesser degrees of success. The standard model of scientific inquiry holds that scholars and scientists in any field ought to restrict themselves to making hypotheses that can then be subject to experimental falsification (Popper, 2008). While there are clear barriers to using this method to study unprecedented future events, it is still possible for ETR researchers to make predictions about issues that are related or adjacent to ETR, or that would alter their subjective probability estimate for an ETR, but would not require humanity going extinct in order to adequately test them. For instance, as a proxy for demographic catastrophe stemming from population growth and resource insufficiency, Paul Ehrlich famously bet the economist Julian Simon that the value of copper, chromium, nickel, tin, and tungsten would rise between September 1980 and September 1990, while in 2002, Rees proposed a bet that 'A bioterror or bioerror will lead to one million casualties in a single event within a six month period starting no later than Dec 31 2020' — a wager which was accepted by Stephen Pinker (Rees and Pinker, 2017). There is clearly epistemic value in making and resolving well-operationalised predictions, and fairly updating our beliefs on their outcomes, if only to keep us from selectively latching onto our predictions and bets only when they are right, while always moving the goalposts when they do not quite come out right on the rationalisation that we were somehow 'right in spirit'. Different approaches to when or where (quantitative) predictions are warranted, and how to treat their outcomes, represent a potential point of epistemic tension between different researchers in this space.

Another approach to scientifically studying extreme technological risk can be taken from engineering and risk assessment tools, such as 'fault trees'. These seek to assess risks that are unlikely to occur but would be catastrophic if they did by first identifying fail states in the system being studied and then working backwards to identify what parameters would need to hold in order for such a fail state to be achieved. While not necessarily applying the scientific method as such, these approaches aim for a degree of analytical rigour and generalisability that allows them to present an objective assessment of risk beyond that typically discussed in the social sciences and humanities. Still, this method is far from bullet-proof: engineers might incorrectly identify failure states, miss important failure states, incorrectly determine what conditions would cause a failure state to occur, or miss failures that can be caused by the operation of a system as a whole (which is harder to model) rather than the operation of its specific parts. Nevertheless, fault tree risk assessments can be valuable for risk management processes because they provide a systematic way of thinking about risks, how they occur, and how they can be prevented. Specifically, they allow researchers to pinpoint useful intervention points (e.g., at earlier stages or at lower cost), gaps in existing interventions (by mapping intervention options to fault trees), or gaps in knowledge or previously under-appreciated areas of uncertainty (e.g., through expert elicitation). Fault trees can also be augmented by the addition of condi-tional probabilities for each step in the tree given what other steps have occurred, in which case they are known as Bayesian Networks. Finally, fault trees present a view of these risks in ways that are easy to interpret and thus build on or respond to. Some especially well-developed fault trees for thinking about ETR include Barrett and Baum (2016) on AI risk, and Baum *et al.* (2018) on nuclear war risk.

There are many considerations that favour one or more of these approaches, including their transparency and rigour, ease and accessibil-ity, and general level of scientific acceptance. Beard *et al.* (2020) under-take an evaluation of how these tools have been used thus far to assess the level of risk from different existential hazards and argue that there is no clear methodology that is 'best', but that unfortunately many of these approaches have been routinely implemented badly and/or for tasks for which they are less well suited. In some respects, the differences between these methodologies fall along familiar academic fault lines, such as between qualitative and quantitative methods or between the 'hard' and 'social' sciences. However, these apparently clear distinctions should be interpreted with a higher-than-usual degree of caution, given that in almost all instances the study of ETR involves investigating extreme and unprecedented events that are hard to model accurately and for which there is very limited evidence. It may well be that the ultimate desire for a more scientific approach to managing ETR requires not the selection of scientific methods and approaches but rather rigour and transparency in the implementation of whatever method is being used so that it is easy for others to understand its key assumptions, parameters, and areas of uncertainty and make up their own mind about whether these are appropriate or require adjustment.

8.3.5 *Q5. How should we achieve influence and impact? How does this vary from different theories of change?*

In the Managing Extreme Technological Risk programme, it was assumed that researchers would build 'vertical' connections to policymakers, the public, academia, and industry and also via the process of 'embedding', giving individual researchers the opportunity to spend time working away from CSER in the teams with which we wished to engage. This would give us the opportunity to gain greater understanding of how these communities perceived, discussed, and addressed risk, and how we might be able to work with them to enhance this. However, embedding can be hard to achieve because of a need to find genuinely mutually beneficial projects to work on and also because of the significant opportunity costs of lost research time and separation from a rapidly evolving field and intensely interconnected community, both at the Centre and with our close partners. Nevertheless, CSER researchers were able to form productive mutual collaborations with a range of groups including working with technology companies to increase trust in AI development and with politicians to found a parliamentary group for future generations.

Another approach to achieving impact might be roughly termed the 'elite spokesperson' approach. Before discussing this in detail, it is relevant to consider how this is tied up with another key trend in the ETR community: the pre-eminence of private philanthropic funding for research grants and programs.[14] Much attention of late has been given to the significant amounts of funding entering the ETR (and adjacent) communities from high-profile philanthropic foundations, including the Open Philanthropy Project and (prior to its collapse in November 2022) the FTX Future Fund (MacAskill, 2022a). The scandalous collapse of FTX may, amongst its many impacts and victims, have had an unexpected and significant negative impact on the resources available in this space.[15]

---

[14]See, for instance, Cremer and Kemp (2021, p. 1) for an indicative discussion of prominent funding sources in the ETR community.

[15]For an informal estimate, see Sempere (2022).

Nonetheless, the overall role of these types of donors means that parts of the ETR community have become more well-resourced, influential, and expansive than in the past. Yet, it has also resulted in a potentially narrow and idiosyncratic funding base.

Of course, philanthropic funding has been on the rise in supporting many fields of science (Sohn, 2023), with philanthropists providing significant fractions of research funding, especially in the US (Kastner, 2018; Shekhtman *et al.*, 2022; Gouwenberg *et al.*, 2016). As in other fields, philanthropic funding sources can have distinct benefits for some areas of ETR work. Meta-science research suggests that standard academic funding bodies are generally conservative and cautious, prefer work with clear, direct, and near-term applications, and may be ill suited for truly interdisciplinary work on highly uncertain topics (Currie, 2019), with the result that scientific work has become steadily less 'disruptive' over time (Kozlov, 2023). By contrast, philanthropic models of research funding, at their best, are able to pursue a more explorative or 'hits-based' (Karnofsky, 2016) approach to funding early-stage work, landscape assessments, or interventions on problems with high uncertainties. This could make funding models offered by organisations, such as the Open Philanthropy Project, unusually (or at least comparatively) well suited for supporting work on many types of ETR, where there may be only limited historical precedent or base rates to learn from or to be motivated into action by, and where success may require the funding of projects that are uncertain but have high expected value, in that they could have a tremendous positive impact, even if their probability of success is low. Indeed, the ability of philanthropic models to be able to initially fund and pursue not just analysis of ETR but also novel interventions on ETR might be especially worthwhile, as doing so could provide early evidence on these interventions' (in)effectiveness, and so can inform future investment decisions on ETR mitigation (Askell, 2019).

Nonetheless, such philanthropic sources of funding also present challenges: these include (1) the potential shaping of research agendas in ways that may be insular or at least may miss crucial considerations; (2) potential challenges about the inability to easily fund certain interventions (such as intergovernmental institutions like the under-funded Biological Weapons Convention) which may either gain their legitimacy from public funding, or where there are regulatory funding limits on private donations (e.g., around electoral campaigns or lobbying); (3) vulnerability to sudden downside shocks that would simultaneously affect large parts of the funding pool (e.g., a new crypto-bubble or collapse, as will be discussed); (4) the sustained efficacy of the interventions that may risk putting 'all eggs in one basket', standing or falling with the continued accurate judgement of a relatively small group of grantors (or re-grantors), who may simply not be able to sufficiently keep track of relevant developments and scientific insights, across multiple fields, to assess highly variable interventions or grants; finally, (5) the philanthropic model that might also foreclose more foundational critiques of or to the ETR community. For instance, a deep critique of such funding arrangements could draw from Anand Giridharadas's (2018) critique of how elite funding of work that (often very sincerely) purports to be 'world changing' could nevertheless ends up propping up a worldview that is sympathetic to and supportive of certain capitalist interests, and dismissive of the potential need for fundamental systemic change, or the redistribution of power and privilege.[16]

More practically, the dependency on an increasingly concentrated base of philanthropic funding has introduced additional risks for the ETR community. For one, there is a risk that this funding may be unevenly distributed, with significant funding going to only some types of existential hazards.[17] CSER's work, by contrast, focuses more on academic research and considers a far wider range of risks and approaches to the study and mitigation of risk, but remains predominantly funded by other philanthropic sources, such as the Templeton World Charity Foundation (who funded the Managing Extreme Technological Risk programme), Grantham Foundation, and Isaac Newton Trust, where it must continue to compete against other important kinds of work. Secondly, as illustrated by the significant fallout of the financial implosion of FTX and the resulting collapse of its Future Fund, a highly concentrated philanthropic funding base that operates without crucial oversights or institutions introduces

---

[16] However, for a recent response to such 'hyper-criticism' of philanthropy, see also Breeze (2021), arguing that while not perfect, philanthropy continues to offer significant value distinct from government and market.

[17] However, it is unclear how strong this discrepancy is: for instance, a 2021 overview (based on data from the Open Philanthropy Project for 2019) suggested that the breakdown of funding aligned to the effective altruist movement was global health (44%), farm animal welfare (13%), biosecurity (10%), AI risks (10%), near-term US policy (8%), and 'Effective Altruism/rationality/cause prioritisation' (6%) (Todd, 2021). An informal but more up-to-date overview is given by Sempere (2022).

catastrophic risks to the ETR community itself. Both factors therefore argue in favour of a significant need for diversification.

Of course, beyond the prominent position that high-profile philanthropic funding bodies hold, the ETR community may have many other potential interactions or avenues to work with influential actors, such as prominent academics, developers of emerging technologies, and politicians and civil servants, to name but a few.

Engaging with high-visibility influential actors can be a natural point of entry for seeking to exert influence over ETR mitigation measures. However, this is not just because they may be able to exert significant direct leverage on key policies but also because they can be key public 'spokespeople', with the potential platform and considerable symbolic capital to be able to strengthen ETR (either particular sources, or as a class) as credible threats to civilisation, demanding and deserving response. However, the relationship between influential elite actors and ETR can be complex. This is particularly the case where it comes to very visible 'spokespersons'. Take one prominent, if extreme, example, Elon Musk: while his interests have shifted over time, he has been a proactive and early supporter of many causes in the ETR community, and has also invested heavily in technologies (e.g., batteries and electric vehicles) that could help reduce some global risks around climate and critical resources. At the same time, his idiosyncratic and eclectic approach to innovation and technology management has arguably stoked a competitive racing mentality in some ETR-relevant industries (notably AI), while his approach to other global risks (such as the COVID-19 pandemic) has been dismissive (Walsh, 2021). While the support and influence of particular types of elite actors may be generally valuable or necessary for raising awareness about or making progress against some risks, it is nevertheless clear that at the very least, the ETR community needs to involve a diverse range of stakeholders and be willing to distance itself from even its keenest supporters if they advocate (elsewhere) for policies or politics that plausibly contribute to ETR. At the same time, the community should be cautious about deploying such steps, and ensure these are not used to settle any possible political or personal disagreements amongst community members, within what should be, or might need to be, a broad-tent community working towards the shared purpose of mitigating ETR.

There are additional reasons why the ETR field may want to reach beyond a focus on influencing a few powerful actors. For instance, as Catherine Rhodes points out in Chapter 6, it is unlikely that any single measure to protect against ETR that they might have influence over would be sufficient to prevent the development or use of potentially dangerous technologies. Rather, we need a 'we(b) of prevention' made up of 'multiple layers and components that, functioning together, greatly improve the strength of such activities'. Indeed, even such a diversified approach may not be enough to achieve technological safety since, as Dr Rhodes goes on to argue, ensuring that science is conducted responsibly may require having 'scientific communities and individual scientists attuned to ethical responsibilities related to their work and recognising the key role that value judgements play in science'. Hence, to succeed, the audience for our engagement activities may truly need to extend to the entire scientific community, and not simply involve those who we judge to have significant influence over a risk.

Similarly, in Chapter 7, Haydn Belfield points out the power of activism as a powerful tool within the AI development community, whereby strong norms in favour of AI safety are being developed from the bottom up, rather than imposed by the most powerful. He argues that the relatively small, narrow talent pool gives that community an unusual degree of economic leverage and that it has used strong shared culture and institutional enabling structures to harness this power to have meaningful impact on corporate behaviour and the kinds of technologies that are being developed. While other types of ETR may be emerging in fields that are less well suited to such activism, this case study seems to suggest the possibility for more empowering approaches to ETR mitigation that focus on helping individuals organise collectively in defence of our shared future, rather than trying to impose a centralised vision of safety from the top down.

In Chapter 2, Adrian Currie and Lalitha Sundaram point to a very different model for engagement between the science of ETR and Policy. 'What needs to percolate [from science to policy]', they argue, is:

> the idea that there are many ways in which the future might unfold, and what needs instilling [in the minds of policymakers] is a curiosity about (and wariness concerning!) those futures. When framed as a piece of traditional policy advice, "take the long term into consideration when making policy" seems rather toothless, vague and unsatisfactory but when seen as a constant drip-feed from multiple disciplines across months and years that gradually "enlightens" those involved in policymaking, it could be quite powerful and more long-lasting in its impact.

Furthermore, they argue that 'this "drip feeding" model isn't really unidirectional' and that it is compatible with an alternative model from science and technology studies — co-production. They define this using Jasanoff's description of the relationship between science and policy in which '[s]cience in the co-productionist framework is understood as neither a simple reflection of the truth about nature nor an epiphenomenon of social and political interests' (2004). This points to a complex bidirectional relationship where governance and policy are informed and shaped by science but where political forces also shape research. They note that:

> Importantly, under this framework science and its processes are not seen as simply tools with which to solve problems: it is recognised — as in the study of extreme technological risks — that science and technology can be a potential source of problems for governance and policy to grapple with … [T]his organising theory of co-production has at its core four areas where co-production manifests, that we suggest in this chapter will be useful for existential risk studies to attend to: the emergence of new phenomena (for us, of ideas), disagreement surrounding those ideas, standardisation in how they are researched and promulgated, and enculturation in how research is performed.

This model has strong parallels with the more direct approach to achieving impact and policy change developed for METR's successor programme, ASOGR, in which participatory foresight, active communications and engagement, and policy co-creation and design were used to develop strong two-way interconnections between CSER researchers and those we wished to influence. Examples of this have been the establishment of a Science–Policy Interface for Global Catastrophic Risk, involving existential risk researchers and diplomats in dialogue about policy priorities and opportunities for change, and the increasing use of scenario-based tools, like Intelligence Rising and, more recently, Participatory Evolution (https://parevo.org/) as an opportunity to simultaneously inform people about the nature of the risks that we face and study their reaction to these risks and what may help improve future responses. The key implications for the ETR community are to avoid thinking of ourselves as outsiders with special knowledge whose job is merely to communicate that to a policy audience and more as engaged scientific citizens. Even if our goal is merely to undertake outside working, where we analyse problems from an objective, or even distanced, point of view, we still need to engage in dialogue with real-world decision-makers to understand what problems we are actually trying to solve and to create different pathways for turning our knowledge into impact. However, if we are more orientated towards inside working, then the need for such two-way engagement becomes even stronger as we seek to reduce the gap between practitioners and risk researchers so as to entrench an understanding of ETR throughout the decision-making process.

8.3.6 *Q6. How might greater diversity benefit our individual work and the management of ETR more generally?*

The Managing Extreme Technological Risk programme deliberately brought in researchers from a range of disciplines to demonstrate how scholars working on diverse components of understanding and managing ETR could work together on a common endeavour of applying, learning, and refining its model of a scientific approach to managing ETR. Within its initial research phase, and expanded since, CSER thus had a diverse range of disciplines within its staff working both to apply their own knowledge and understanding to particular problems, and to combine their insights in collaborative projects and develop and refine their own techniques and methodologies. Interdisciplinary working is crucial for effectively understanding and addressing existential risks and other major challenges that humanity collectively faces. Genuine collaborative efforts and combined approaches are extremely productive, but can be challenging to achieve. Even disciplines that appear quite 'close', e.g., different social sciences, utilise different 'languages', favour particular approaches, and have distinct underlying assumptions that may well not be made explicit. Effective interdisciplinary working can therefore be challenging to navigate and subject to mistranslation and misunderstanding. The epistemic humility pointed to in earlier chapters is important here.

However, disciplinary diversity is clearly not the only kind of diversity that can benefit our understanding and management of ETR. Differences in the background, life experience, personal networks, and expected future trajectories of scholars all have a significant impact upon the work that they do. Thus, the same arguments that run in favour of interdisciplinarity in the field would also seem to support the idea that it should be diverse and inclusive in other respects as well. Indeed, we have often seen how diversity, of many forms, is key to addressing a large number of challenges faced by the ETR community — diversity of motivations, diversity of prioritisation and selection of research topics, diversity of philosophical approaches and theories of change, diversity of resourcing, diversity in audiences engaged for impact and so on.

Given how small the field of researchers dedicated to managing ETR is, it might seem that it should not be difficult to gauge how homogeneous it is in terms of the scholars within it, their backgrounds, characteristics, and other factors. However, in practice, managing ETR is also a highly fluid field, at the nexus of many overlapping communities and academic fields. Its many organisations, hosting a wide range of visitors, collaborators, and affiliates, may or may not identify (or be identified) as Existential Risk Studies or other fields explicitly tied to managing ETR. This complicates any 'census' of the ETR community. To our knowledge, there has only been one, relatively informal and non-comprehensive study in this domain, which looked at demographic characteristics in the institutional subsection of the ETR community, which is associated with, and supported by, the Effective Altruism movement (Anonymous EA Forum Account, 2020). This assessment found that only 13–14% of researchers in this space were people of colour (i.e., non-white), with almost half of ETR institutions surveyed (10/21) having only white members in their core staff and a quarter (5/21) having only white members across all staff. It should be noted that there appears to be a significant difference between university-affiliated institutions (where 20% of staff are non-white) and those unaffiliated to universities (where only 3% are). There were also disparities among different roles, with more people of colour working in junior research and support roles and fewer in senior leadership positions. People of colour were, regrettably but unsurprisingly, most represented among volunteer, part-time, freelance, and internship positions where they made up 29% of all staff.

Other surveys indicate similar unbalances with regards to (binary) gender diversity, with 71% of respondents to a 2019 Effective Altruism survey reporting their gender as male and 26.9% as female (Dullaghan, 2019b). This lack of gender balance, which could be even worse in other parts of the ETR community that have not kept data, is a challenge. It may be a problem for managing ETR, not just for intrinsic normative reasons, but also if one takes a purely instrumental perspective on ETR mitigation. For instance, Barnhart (2022) has argued that those concerned about the existential risks of AI technology should advocate for the greater promotion of gender diversity amongst AI researchers, since polling suggests that women researchers are significantly more likely to think that issues of AI safety and AI governance research should be prioritised. In this way, the greater inclusion of women, both in decision-making roles in ETR-sensitive fields as well as in the ETR community itself, could be a reliable and key intervention.

One area in which the ETR community stands out as potentially more diverse than wider society is in relation to neurodiversity, mental health, and potentially other forms of disability. While no formal assessment of this in the community has been made, a reader survey of Slate Star Codex, a popular rationalist blog with some links to the ETR community, found that '[s]lightly less than a third of Slate Star Codex readers have anxiety; slightly more than a third have depression. Just under a fourth [...] are on the Autism spectrum (or believe themselves to be)'. More tentatively, there is also some evidence that the ETR community also has a greater than expected level of sexual and gender diversity, with the same survey finding that '5.5% of [...] readers are transgender or genderqueer, [compared] to the 0.3% of American adults who identify as transgender'. Anecdotally, we have found similar trends in the wider community and especially amongst younger and more junior researchers. It is notable, however, that while a community likely gains substantially from having a diversity of ways of thinking and feeling about the world, this fact is often not mentioned, let alone celebrated. A greater understanding of neurodiversity and gender and sexual diversity, as well as reflection on their benefits, could be one way of opening up other conversations about the need for a plurality of voices and perspectives, as well as making the field more open and accessible to everyone.

Beyond a challenge for talent, increasing the diversity of the field could have important and critical effects for helping manage ETR effectively. As we discussed in Question 1, there are important implications for priority setting in the field, and the current set of approaches, in isolation, could leave the field intellectually impoverished, with an overreliance on a single set of assumptions about the world that are potentially flawed and certainly contested. At the same time, it should be noted that this kind of overreliance is neither universal nor static, and there are proactive efforts being made in many areas of ETR to diversify across many dimensions. It would therefore be beneficial to the ETR community, as one that aims to represent and safeguard the future of humanity as a whole, to pay further attention to examining some of our core assumptions (and those of antecedent disciplines) and to further increase the diversity of disciplinary backgrounds engaged in the field, particularly those that have a more 'critical' understanding of their fields. Such work can draw on contemporary discourses on decolonisation and diversification, and communicate this better to new entrants in the field.

To give some potential examples (which would require much further elaboration): debates over how ETR researchers should view the macro-strategic and ethical implications of technological interventions to mitigate or stabilise ecosystem collapse ('techno-adaptation') could draw important insights from the field of 'green transhumanism' (Holt, 2021), while arguments over the value of the long-term future can draw on complementary ecocentric arguments for space expansion ('greening the universe') (Owe, 2022). Similarly, the discipline of heterodox economics draws on models of steady-state economics: if these could be shown to be sufficiently stable over longer-term timelines, such strategies could offer complementary or alternate avenues to long-term civilisational well-being that could be more reliable and tractable than accounts that pursue 'sustainability as trajectory' (Bostrom, 2013).

Existential Risk Studies and the ETR community could therefore have an important role to play in critically questioning or even challenging overtly linear views of human progress, improvement, or perfection or even developing entirely new approaches that are more able to embrace the whole of the human experience. Although challenging to some, such discussions can be both creative and enlightening, especially as they enable the cross-pollination of ideas as well as the opportunity to invite new people into discussions about ETR.

8.3.7 *Q7. How can we most constructively reflect on such questions as the field continues to grow?*

One key theme from the Managing Extreme Technological Risk programme is the need for responsible science. However, as has already been pointed out in the introduction and Chapter 6, this would seem to beg the question: what is a responsible science of managing ETR? This is clearly a question with many facets; however, we would like to end this section by considering just one of these, which we feel has particular relevance to how constructive reflection might help this urgent and important project to move forwards.

All fields of enquiry involve uncertainty — after all, there would be no point in research otherwise. However, for managing ETR, with all the challenges and issues we have here described, uncertainty is of particular and perhaps obvious importance. This is because we do not have to worry about simply communicating uncertainty (i.e., telling people our level of certainty in a given prediction or that a given strategy will be worthwhile), but about communicating uncertainty about uncertainty or urgency under uncertainty (i.e., convincing people of the need to take action at all, while also being honest about how difficult it is to know what the problem is or how to solve it). Our best estimates or models when it comes to the nature of ETR may well be the best of anyone's guesses, but they are sometimes still far from meeting conventional scientific standards of proof and face a number of challenging critiques.

On the one hand then, many ETR researchers take their responsibilities of informing, advising, and even influencing very seriously and thus need to develop tools for communicating to these audiences appropriately. On the other hand, the field does remain a scholarly/academic one, where, at least ideally, caution and transparency about the limitations of our methods and approaches are to be encouraged, and where concepts of falsifiability, objectivity, and impartiality are at least to be aspired to. Recognising this is itself a key challenge, and the field is also evolving in this respect, moving from purely theoretical considerations to methodological ones. As exemplified in Managing Extreme Technological Risk — the subject of this volume — scholars do adapt, refine, and build on concepts and tools of 'scientificness' in ways that are appropriate to the field.

One possible reaction here might be to propose differential communication strategies for each of these purposes and audiences. Following such an approach, the background research that consists of in-the-weeds and nuanced discussions about evidence, uncertainty, and disagreement would be mostly reserved for the academic debates, while 'take home messages' and high-level recommendations and strategies would be distilled and published separately for public and policymaker audiences in a way that downplays some of the uncertainty and complexity for purposes of message clarity. However, there are risks to such a segmented communication strategy: not only would it feel somewhat intellectually patronising but it might not work as intended — the underlying uncertainties are likely to be 'found out' — and if so, this could easily backfire, by (legitimately) drawing into question our credibility, honesty, and independence as researchers, no matter how benign our intentions.

Another, better approach is the more intellectually honest one: making clear our uncertainty in as accessible a manner as we can, while nevertheless stressing the importance of tackling ETR. For these cases, and these communities, it might be most appropriate to reflect openly on object-level uncertainties, while simultaneously conveying the underlying message that ETR still needs urgent attention and precautionary action: to stress that while we are uncertain (and may prove mistaken) about the specifics of a particular ETR, we can be confident about the aggregate importance and urgency of ETR as an overarching human challenge.

What remains then is another purpose of communication, within the field and externally: about how the ETR community itself should reflect on its own topic prioritisation (in terms of frameworks, disciplinary lenses, or funding), approaches, methods, theories of change, and ways of pursuing impact, composition, and future trajectory. Of course, this process of reflection about these questions will be a continuation of existing conversations within the field, and will be a long-term endeavour. When we engage in these efforts, it will be important to recognise our uncertainties about these questions and their 'right' answers, while also understanding that that uncertainty is not a sufficient reason to delay or avoid these conversations.

8.4 Conclusion

In this chapter, we have asked seven key questions about managing ETR and sought to address these through our experience of the Managing Extreme Technological Risk programme and the wider ETR community; however, we have not attempted to provide definitive answers to any of them. To do so would miss the point. These are hard problems that need careful and ongoing consideration, and that consideration should start with an appreciation that there really are many different ways of answering these questions, each with advantages and drawbacks.

One of the reasons that we have chosen to publicly pose and consider these questions now is that they are rooted in our own lived experiences of being researchers in this community of alignment. The Managing Extreme Technological Risk programme adopted an experimental approach to bringing together researchers from diverse disciplines to build understanding of several aspects of managing ETR, and explicitly encouraged them to do so with connection to and in collaboration with policy and scientific communities. As an emerging and quite rapidly developing field of study, the ETR community is, unsurprisingly, relatively small (at least in terms of pure researchers and established institutions) and has some 'founder effects'. Nonetheless, there is already some diversity in relation to, for example, theories of change, areas of focus, and methodological preferences. This diversity is likely to continue to grow, although there may also be bounds to this if Existential Risk Studies and the wider ETR community are to remain coherent and aligned.

Our hope must be that as the field grows larger and more established, this diversity and pluralism will increase, and that no single approach or theory will come to dominate or lead to greater homogenisation. Crucially, there are also several ways in which the ETR community can benefit from reflecting on these questions in a way that allows for more pluralism and diversity, both in the disciplinary backgrounds that can find a 'home' within the field as well as in terms of who is attracted to the field, welcomed within it, and supported in their participation and contribution. The only thing that would be worse than seeing this difference of opinions collapse within the field would be to find that people stop asking these questions at all because they become too controversial or difficult, or because the answers seem too uncertain. That would be a bad day indeed for this field and, given the importance of what we work on, potentially for humanity as a whole.

Bibliography

Aaken, A. V. (2016). Is international law conducive to preventing looming disasters? *Global Policy,* S1(7), 81–96. https://doi.org/10.1111/1758-5899.12303.

Abdalla, M. and Abdalla, M. (2021). The Grey Hoodie project: Big tobacco, big tech, and the threat on academic integrity. In *Proceedings of AIES 2021 — Proceedings of AAAI/ACM Conference on AI Ethics Society*, pp. 287–297. https://doi.org/10.1145/3461702.3462563.

Aird, M. (2020). Differential progress/intellectual progress/technological development. *Effective Altruism Forum*. Available at: https://forum.effectivealtruism.org/posts/XCwNigouP88qhhei2/differential-progress-intellectual-progress-technological.

Aird, M. (2021a). What's wrong with the EA-aligned research pipeline? *Effective Altruism Forum*. Available at: https://forum.effectivealtruism.org/posts/4wk7GZpuBXufTiKbh/what-s-wrong-with-the-ea-aligned-research-pipeline.

Aird, M. (2021b). Thoughts on "a case against strong longtermism" (Masrani). *Effective Altruism Forum*. Available at: https://forum.effectivealtruism.org/posts/EDq4GLD67yAnWP6oN/thoughts-on-a-case-against-strong-longtermism-masrani.

Alexander, S. (2020). A failure, but not of prediction. *Slate Star Codex*. Available at: https://slatestarcodex.com/2020/04/14/a-failure-but-not-of-prediction/.

Alexander, S. (2022). "Long-termism" vs. "existential risk". *Effective Altruism Forum*. Available at: https://forum.effectivealtruism.org/posts/KDjEogAq WNTdddF9g/long-termism-vs-existential-risk.

Anonymous EA Forum Account. (2020). Racial demographics at longtermist organizations. *Effective Altruism Forum*. Available at: https://forum. effectivealtruism.org/posts/W8S3EuYDWYHQxm77u/racial-demographics-at-longtermist-organizations.

Aschenbrenner, L. (2019). Existential risk and growth. *GPI Working Paper*. Global Priorities Institute, University of Oxford. Available at: https:// globalprioritiesinstitute.org/wp-content/uploads/Leopold-Aschenbrenner_ Existential-risk-and-growth_.pdf.

Aschenbrenner, L. (2021). Burkean Longtermism. *For Our Posterity* (blog). Available at: https://www.forourposterity.com/burkean-longtermism/.

Askell, A. (2019). Chapter 3 evidence neutrality and the moral value of information. In H. Greaves and T. Pummer (eds.), *Effective Altruism: Philosophical Issues*. Oxford: Oxford University Press.

Avin, S. and Amadae, S. M. (2019). Autonomy and machine learning at the interface of nuclear weapons, computers and people. In V. Boulanin (ed.), *The Impact of Artificial Intelligence on Strategic Stability and Nuclear Risk*. Stockholm International Peace Research Institute. https://doi.org/10.17863/CAM.44758.

Avin, S., Gruetzemacher, R., and Fox, J. (2020). Exploring AI futures through role play. In *Proceedings of AIES 2020 — Proceedings AAAI/ACM Conference on AI Ethics Society*, pp. 8–14. https://doi.org/10.1145/3375627.3375817.

Avin, S., Sundaram, L., Whittlestone, J., Maas, M., and Hobson, T. (2021). Submission of evidence to the house of lords select committee on risk assessment and risk planning. Report. Available at: https://doi.org/10.17863/ CAM.64180.

Avin, S., Wintle, B. C., Weitzdörfer, J., Ó hÉigeartaigh, S. S., Sutherland, W. J., and Rees, M. J. (2018). Classifying Global Catastrophic Risks. *Futures*, 102, 20–26. https://doi.org/10.1016/j.futures.2018.02.001.

Ballantyne, N. and Ditto, P. H. (2021). Hanlon's Razor. *Midwest. Stud. Philos*. Available at: https://philpapers.org/archive/BALHR.pdf.

Balwit, A. (2021). Response to recent criticisms of Longtermism. *Effective Altruism Forum*. Available at: https://forum.effectivealtruism.org/posts/ kageSSDLSMpuwkPKK/response-to-recent-criticisms-of-longtermism-1.

Barnhart, J. (2022). Gender diversity may be in longtermists' interests. Unpublished draft.

Barrett, A. and Baum, S. (2016). A model of pathways to artificial superintelligence catastrophe for risk and decision analysis. *J. Exp. Theor. Artif. In.*, 29, 1–18.

Barrett, S. (2007). *Why Cooperate?: The Incentive to Supply Global Public Goods*. Oxford: Oxford University Press.

Baum, S. D. (2015). The far future argument for confronting catastrophic threats to humanity: Practical significance and alternatives. *Futures*, 72, 86–96.

Baum, S. D. (2020). Quantifying the probability of existential catastrophe: A reply to Beard *et al. Futures*, 123, 102608.

Baum, S. D. and Handoh, I. C. (2014). Integrating the planetary boundaries and global catastrophic risk paradigms. *Ecol. Econ.*, 107, 13–21.

Baum, S. D., Barrett, T., and Fitzgerald, M. (2022). GCRI statement on pluralism in the field of global catastrophic risk. Global Catastrophic Risk Institute. Available at: https://gcrinstitute.org/gcri-statement-on-pluralism/.

Baum, S. D., de Neufville, R. and Barrett, A. (2018). A model for the probability of nuclear war. *Global Catastrophic Risk Institute Working Paper*. Available at: https://papers.ssrn.com/sol3/papers.cfm?abstract_id=3137081.

Baum, S. D., *et al.* (2019). Long-term trajectories of human civilization. *Foresight*, 21, 53–83.

Beard, S. J. and Bronson, R. (2023). A brief history of existential risk and the people who worked to mitigate it. In S. J. Beard *et al.* (eds.), *The Era of Global Risk: An Introduction to Existential Risk Studies* (pp. 1–26). Cambridge: Open Book Publishers.

Beard, S. J. and Torres, P. (2020). Ripples on the Great Sea of Life: A Brief History of Existential Risk Studies. Available at SSRN 3730000.

Beard, S. J., Rowe, T., and Fox, J. (2020). An analysis and evaluation of methods currently used to quantify the likelihood of existential hazards. *Futures*, 115, 102469.

Beard, S. J. *et al.* (2021). Assessing Climate Change's Contribution to Global Catastrophic Risk. *Futures*. 127, 102673.

Beckstead, N. (2013). On the overwhelming importance of shaping the far future. Doctoral Thesis, Rutgers University.

Beckstead, N. (2015). Differential technological development: Some early thinking. *The GiveWell Blog*. Available at: http://blog.givewell.org/2015/09/30/differential-technological-development-some-early-thinking/.

Beckstead, N., Aschenbrenner, L., MacAskill, W., and Ramakrishnan, K. (2022). Announcing the future fund. *Effective Altruism Forum*. Available at: https://forum.effectivealtruism.org/posts/2mx6xrDrwiEKzfgks/announcing-the-future-fund-1.

Belfield, H. (2020). Activism by the AI community: Analysing recent achievements and future Prospects. In *Proceedings of AAAI/ACM Conference on AI, Ethics, and Society.* pp. 15–21. https://doi.org/10.1145/3375627.3375814.

Belfield, H. (2023). Collapse, recovery and existential risk. In M. Centeno, P. Callahan, P. Larcey, and T. Patterson (eds.), *How Worlds Collapse: What*

History, Systems, and Complexity Can Teach Us About Our Modern World and Fragile Future (pp. 61–92). New York: Routledge Press.

Benquo. (2017). OpenAI makes humanity less safe. *LessWrong*. Available at: https://www.lesswrong.com/posts/Nqn2tkAHbejXTDKuW/openai-makes-humanity-less-safe.

Boin, A., *et al.* (2016). *The Politics of Crisis Management: Public Leadership under Pressure* (2nd edn.). Cambridge: Cambridge University Press. https://doi.org/10.1017/9781316339756.

Bostrom, N. (2002). Existential risks: Analyzing human extinction scenarios and related hazards. *J. Evol. Technol.*, 9(1).

Bostrom, N. (2003). Astronomical waste: The opportunity cost of delayed technological development. *Utilitas*, 15, 308–314.

Bostrom, N. (2011). Information hazards: A typology of potential harms from knowledge. *Rev. Contemp. Philos.*, 10, 44–79.

Bostrom, N. (2013). Existential risk prevention as a global priority. *Global Policy*, 4, 15–31.

Bostrom, N. (2014a). Crucial considerations and wise philanthropy. In *Presented at the Good Done Right Conference*, All Souls College, Oxford. Available at: https://nickbostrom.com/lectures/crucial_final.pdf.

Bostrom, N. (2014b). *Superintelligence: Paths, Dangers, Strategies*. Oxford: Oxford University Press.

Bostrom, N. (2019). The Vulnerable World Hypothesis. *Global Policy*, 10(4), 455–476. https://doi.org/10.1111/1758-5899.12718.

Bostrom, N. and Circovic, M. M. (2011). Introduction. In N. Bostrom and M. M. Circovic (eds.), *Global Catastrophic Risks*. Oxford: Oxford University Press.

Bostrom, N., Dafoe, A., and Flynn, C. (2019). Public policy and superintelligent AI: A vector field approach. In S. M. Liao (ed.), *Ethics of Artificial Intelligence*. Oxford: Oxford University Press.

Botts, T. F., *et al.* (2014). What is the state of blacks in philosophy? *Crit. Philos. Race.*, 2, 224–242.

Breeze, B. (2021). *In Defence of Philanthropy*. Columbia University Press.

Brundage, M., *et al.* (2018). The malicious use of artificial intelligence: Forecasting, prevention, and mitigation. *arXiv preprint*. arXiv:1802.07228.

Brundage, M., *et al.* (2020). Trustworthy AI development: Mechanisms for supporting verifiable claims. *arXiv preprint*. arXiv:2004.07213.

Buehler, B. and Kessler, A. (2016). The ideology trap. Available at: http://www.sfu.ca/~akessler/wp/ideologues.pdf.

Carpenter, C. (2014). *Lost Causes: Agenda Vetting in Global Issue Networks and the Shaping of Human Security*. Cornell University Press.

Carson, R. (1962) *Silent Spring*. Houghton Mifflin.

Cave, S. and ÓhÉigeartaigh, S. S. (2018). An AI race for strategic advantage: Rhetoric and risks. In *Proceedings of the 2018 AAAI/ACM Conference on AI, Ethics, and Society*, pp. 36–40.

Chivers, T. (2019). *The AI Does Not Hate You: Superintelligence, Rationality, and the Race to Save The World*. London: Weidenfeld & Nicolson.

Christian, B. (2020). *The Alignment Problem: Machine Learning and Human Values*. New York: W. W. Norton & Company.

Clarke, S., Carlier, A. and Schuett, J. (2021). Survey on AI existential risk scenarios. *Effective Altruism Forum*. Available at: https://forum.effectivealtruism. org/posts/2tumunFmjBuXdfF2F/survey-on-ai-existential-risk-scenarios-1.

Colvin, R. M., *et al.* (2020). Learning from the climate change debate to avoid polarisation on negative emissions. *Environ. Commun.*, 14, 23–35.

Crawford, J. (2021). Help me find the crux between EA/XR and progress studies. *Effective Altruism Forum*. Available at: https://forum.effectivealtruism.org/ posts/hkKJF5qkJABRhGEgF/help-me-find-the-crux-between-ea-xr-and- progress-studies.

Cremer, C. Z. (2020). Objections to value-alignment between effective altruists. *Effective Altruism Forum*. Available at: https://forum.effectivealtruism.org/ posts/DxfpGi9hwvwLCf5iQ/objections-to-value-alignment-between- effective-altruists.

Cremer, C. Z. and Kemp, L. (2021). Democratising risk: In search of a methodol- ogy to study existential risk. SSRN. Available at: https://papers.ssrn.com/ sol3/papers.cfm?abstract_id=3995225.

Cremer, C. Z. and Whittlestone, J. (2021). Artificial canaries: Early warning signs for anticipatory and democratic governance of AI. *Int. J. Interact. Multimed. Artif. Intell.*, 6(5), 100–109. doi: 10.9781/ijimai.2021.02.011.

Cremer, C. Z., *et al.* (2021). Kill the bill to save the future: How the policing & crime bill for England and Wales makes us more vulnerable to catastrophe. *Medium*. Available at: https://medium.com/@t.hobson/kill-the-bill-to-save- the-future-e62689e02328.

Critch, A. (2021). Power dynamics as a blind spot or blurry spot in our collective world-modeling, especially around AI. *LessWrong*. Available at: https:// www.lesswrong.com/posts/WjsyEBHgSstgfXTvm/power-dynamics-as- a-blind-spot-or-blurry-spot-in-our.

Critch, A. and Krueger, D. (2020). AI research considerations for human existen- tial safety (ARCHES). Available at: http://acritch.com/arches/.

Currie, A. (2018). Geoengineering tensions. *Futures*, 102, 78–88.

Currie, A. (2019). Existential risk, creativity & well-adapted science. *Stud. Hist. Philos. Sci. A.*, 76, 39–48.

Dafoe, A. (2015). On technological determinism: A typology, scope conditions, and a mechanism. *Sci. Technol. Hum. Values*, 40, 1047–1076. https://doi. org/10.1177/0162243915579283.

Daniel, M. (2021). Progress studies vs. longtermist EA: Some differences. *Effective Altruism Forum*. Available at: https://forum.effectivealtruism.org/posts/eFQoe4CBCYbotDFQp/progress-studies-vs-longtermist-ea-some-differences.

Dasgupta, P. (2019). *Time and the Generations: Population Ethics for a Diminishing Planet*. Columbia University Press.

Davies, R. (2022). ParEvo: Enabling the participatory exploration of alternative futures. Available at SSRN 4261400.

deluks917. (2020). How dependent is the Effective Altruism movement on Dustin Moskovitz and Cari Tuna? *Effective Altruism Forum*. Available at: https://forum.effectivealtruism.org/posts/4BJSXH9ho4eYNT73P/how-dependent-is-the-effective-altruism-movement-on-dustin.

Deudney, D. (2018). Turbo change: Accelerating technological disruption, planetary geopolitics, and architectonic metaphors. *Int. Stud. Rev.*, 20, 221–231. https://doi.org/10.1093/isr/viy033.

Dickens, M. (2016). Evaluation frameworks (or: when importance/neglectedness/tractability doesn't apply). *Philosophical Multicore*. Available at: https://mdickens.me/2016/06/10/evaluation_frameworks_(or-_when_scale-neglectedness-tractability_doesn't_apply)/.

Drexler, E. (1986). *Engines of Creation: The Coming Era of Nanotechnology*. New York: Double Day.

Dullaghan, N. (2019a). EA survey 2018: How welcoming is EA? *ReThink Priorities*. Available at: https://rethinkpriorities.org/publications/eas2018-how-welcoming-is-ea.

Dullaghan, N. (2019b). EA survey 2019: Community demographics & characteristics. *ReThink Priorities*. Available at: https://rethinkpriorities.org/publications/eas2019-community-demographics-characteristics.

Ellen, R. (2021). Some blindspots in rationality and Effective Altruism. *LessWrong*. Available at: https://www.lesswrong.com/posts/Aq4KNxKscywt3yXqk/some-blindspots-in-rationality-and-effective-altruism.

Flink, T. and David K. (2018). The new production of legitimacy: STI policy discourses beyond the contract metaphor. *Res. Policy*, 47, 14–22. https://doi.org/10.1016/j.respol.2017.09.008.

FTX Future Fund. (2022a). Areas of interest. Available at: https://ftxfuturefund.org/area-of-interest/.

FTX Future Fund. (2022b). Principles. Available at: https://ftxfuturefund.org/principles/.

Gabriel, I. (2020). Artificial intelligence, values, and alignment. *Minds Mach.*, 30, 411–437. https://doi.org/10.1007/s11023-020-09539-2.

Gabriel, I. and Ghazavi, V. (2021). The challenge of value alignment: From fairer algorithms to AI safety. *arXiv preprint*. arXiv:2101.06060. Available at: http://arxiv.org/abs/2101.06060.

Giridharadas, A. (2018). *Winners Take All: The Elite Charade of Changing the World*. New York: Alfred A. Knopf.

Global Challenges Foundation / Novus. (2020). *Global Catastrophic Risks and International Collaboration: Opinion Poll 2020*. Global Challenges Foundation. Available at: https://globalchallenges.org/library/survey-global-catastrophic-risks-and-international-collaboration-2020/ (Accessed 30 January 2024).

Good, I. J. (1964). Speculations concerning the first ultraintelligent machine. In F. L. Alt and M. Rubinoff (eds.), *Advances in Computers*. New York: Academic Press.

Gouwenberg, B., *et al.* (2016). Foundations supporting research and innovation in Europe: Results and lessons from the Eufori Study. *Found. Rev.*, 8(1). https://doi.org/10.9707/1944-5660.1287.

Greaves, H. and MacAskill, W. (2019). The case for strong longtermism. *Global Priorities Institute*. Available at: https://globalprioritiesinstitute.org/wp-content/uploads/2019/Greaves_MacAskill_The_Case_for_Strong_Longtermism.pdf.

Greer, T. (2021). The framers and the framed: Notes on the Slate Star Codex controversy. *The Scholar's Stage*. Available at: https://scholars-stage.org/the-framers-and-the-framed-notes-on-the-slate-star-codex-controversy/.

Halstead, J. (2018). Stratospheric aerosol injection research and existential risk. *Futures*, 102, 63–77. https://doi.org/10.1016/j.futures.2018.03.004.

Hanea, A. M., *et al.* (2017). Investigate discuss estimate aggregate for structured expert judgement. *Int. J. Forecast.*, 33, 267–279. https://doi.org/10.1016/j.ijforecast.2016.02.008.

Harth, R. (2021). How to think about and deal with OpenAI. *LessWrong*. Available at: https://www.lesswrong.com/posts/oEC92fNXPj6wxz8dd/how-to-think-about-and-deal-with-openai.

Hendren, S. (2020). Critique or repair? A call to know your post. Sara Hendren. Available at: https://sarahendren.com/2020/06/30/critique-or-repair-a-call-to-know-your-post/.

Ho, P. (ed.). (2017). The challenges of governance in a complex world. *World Scientific*.

Holt, L. A. (2021). Why shouldn't we cut the human-biosphere umbilical cord? *Futures*, 133, 102821. https://doi.org/10.1016/j.futures.2021.102821.

House of Lords Select Committee on Risk Assessment and Risk Planning. (2021). Preparing for extreme risks: Building a resilient society. Available at: https://publications.parliament.uk/pa/ld5802/ldselect/ldrisk/110/110.pdf.

Husbands, J. L. (2018). The challenge of framing for efforts to mitigate the risks of "dual use" research in the life sciences. *Futures*, 102, 104–113. https://doi.org/10.1016/j.futures.2018.03.007.

Irving, G. and Askell, A. (2019). AI safety needs social scientists. *Distill*. Available at: https://distill.pub/2019/safety-needs-social-scientists/.

Issawi, C. P. (1973). *Issawi's Laws of Social Motion*. New York: Hawthorn Books.

Jehn, F. U., *et al.* (2021). Betting on the best case: Higher end warming is under-represented in research. *Environ. Res. Lett.*, 16, 084036. https://doi.org/10.1088/1748-9326/ac13ef.

Joice, W. and Tetlow, A. (2020). Baselines for improving STEM participation: Ethnicity STEM data for students and academic staff in higher education 2007/08 to 2018/19. The Royal Society. Available at: https://royalsociety.org/-/media/policy/Publications/2021/trends-ethnic-minorities-stem/Ethnicity-STEM-data-for-students-and-academic-staff-in-higher-education.pdf.

Kastner, M. (2018). Philanthropy: A critical player in supporting scientific research. Science Philanthropy Alliance. Available at: https://sciencephilanthropyalliance.org.

Karnofsky, H. (2016). Hits-based giving. *Open Philanthropy Project*. Available at: https://www.openphilanthropy.org/blog/hits-based-giving.

Karnofsky, H. (2021a). All possible views about humanity's future are wild. *Cold Takes*. Available at: https://www.cold-takes.com/all-possible-views-about-humanitys-future-are-wild/.

Karnofsky, H. (2021b) Rowing, steering, anchoring, equity, mutiny. *Cold Takes*. Available at: https://www.cold-takes.com/rowing-steering-anchoring-equity-mutiny/.

Kemp, L. (2021). Agents of doom: Who is creating the apocalypse and why. *BBC Future*. Available at: https://www.bbc.com/future/article/20211014-agents-of-doom-who-is-hastening-the-apocalypse-and-why.

Kemp, L., *et al.* (2020). Bioengineering horizon scan 2020. *eLife.*, 9, https://doi.org/10.17863/CAM.52994.

Kemp L., *et al.* (2021). 80 questions for UK biological security. *PLoS ONE*, 16, e0241190.

Klare, M. and Perry, L. (2021). Michael Klare on the Pentagon's view of climate change and the risks of state collapse. *Fut. Life Inst. — Podcast*. Available at: https://futureoflife.org/2021/07/30/michael-klare-on-the-pentagons-view-of-climate-change-and-the-risks-of-state-collapse/.

Kozlov, M. (2023). 'Disruptive' science has declined — And no one knows why. *Nature*. https://doi.org/10.1038/d41586-022-04577-5.

Krakovna, V. (2020). Possible takeaways from the coronavirus pandemic for slow AI takeoff. *AI Alignment Forum* (blog). Available at: https://www.alignmentforum.org/posts/wTKjRFeSjKLDSWyww/possible-takeaways-from-the-coronavirus-pandemic-for-slow-ai.

Kreienkamp, J. and Pegram, T. (2020). Governing complexity: Design principles for the governance of complex global catastrophic risks. *Int. Stud. Rev.*, 3, 779–806. https://doi.org/10.1093/isr/viaa074.

Krepon, M. (2021). *Winning and Losing the Nuclear Peace: The Rise, Demise, and Revival of Arms Control*. Stanford: Stanford University Press.

Krueger, D. (2019). Project proposal: Considerations for trading off capabilities and safety impacts of AI research. *AI Alignment Forum.* Available at: https://www.alignmentforum.org/posts/y5fYPAyKjWePCsq3Y/project-proposal-considerations-for-trading-off-capabilities.

Kuhlemann, K. (2018). Complexity, creeping normalcy and conceit: Sexy and unsexy catastrophic risks. *Foresight*, 21(1).

Kulveit, J. and Gavin. (2022). We can do better than argmax. Effective Altruism Forum. Available at: https://forum.effectivealtruism.org/posts/8Ban7Ano qwdzQphsK/we-can-do-better-than-argmax.

Leech, G. (2018). Existential risk as common cause. Argmin gravitas. Available at: https://www.gleech.org/x-for-all.

Leslie, J. (1996). *The End of the World: The Science and Ethics of Human Extinction.* New York: Routledge.

Lewis, G. (2016). Beware surprising and suspicious convergence. *Effective Altruism Forum.* Available at: https://forum.effectivealtruism.org/posts/omoZDu8ScNbot6kXS/beware-surprising-and-suspicious-convergence.

Lewis, G. (2017). In defence of epistemic modesty. *Effective Altruism Forum.* Available at: https://forum.effectivealtruism.org/posts/WKPd79PESRGZH Q5GY/in-defence-of-epistemic-modesty.

Liu, H.-Y. and Maas, M. M. (2021). 'Solving for X?' Towards a problem-finding framework to ground long-term governance strategies for artificial intelligence. *Futures*, 126, 102672. https://doi.org/10.1016/j.futures.2020.102672.

Liu, H.-Y., Lauta, K. C., and Maas, M. M. (2018). Governing boring apocalypses: A new typology of existential vulnerabilities and exposures for existential risk research. *Futures*, 102, 6–19. https://doi.org/10.1016/j.futures.2018.04.009.

Liu, H.-Y., Lauta, K. C., and Maas, M. M. (2020). Apocalypse Now?: Initial lessons from the Covid-19 pandemic for the governance of existential and global catastrophic risks. *J. Int. Humanit. Leg. Stud.*, 1, 1–16. https://doi.org/10.1163/18781527-01102004.

Maas, M. M. (2018). Regulating for 'normal AI accidents': Operational lessons for the responsible governance of artificial intelligence deployment. In *Proceedings of AIES 2018 — Proceedings of AAAI/ACM Conference on AI Ethics Society*, pp. 223–228. https://doi.org/10.1145/3278721.3278766.

Maas, M. M. (2019). How viable is international arms control for military artificial intelligence? Three lessons from nuclear weapons. *Cont. Sec. Policy*, 40(3), 285–311. https://doi.org/10.1080/13523260.2019.1576464.

Maas, M. M. (2022). A primer & some reflections on recent CSER work (EAB Talk). *EA Forum.* Available at: https://forum.effectivealtruism.org/posts/WJZAc6fTYNbb5DeAW/a-primer-and-some-reflections-on-recent-cser-work-eab-talk.

Maas, M. M., Lucero-Matteucci, K., and Cooke, D. (2023). Military artificial intelligence as a contributor to global catastrophic risk. In S. J. Beard, *et al.* (eds.), *The Era of Global Risk: An Introduction to Existential Risk Studies* (pp. 237–284). Cambridge: Open Book Publishers.

Maas, M. M., *et al.* (2021). Reconfiguring resilience for existential risk: Submission of evidence to the cabinet office on the new UK national resilience strategy. Centre for the Study of Existential Risk. Available at: https://www.cser.ac.uk/resources/reconfiguring-resilience-existential-risk-submission-evidence-cabinet-office-new-uk-national-resilience-strategy/.

MacAskill, W. (2016). Moral progress and cause X. EA global 2016. Effective Altruism. Available at: https://www.effectivealtruism.org/articles/moral-progress-and-cause-x/.

MacAskill, W. (2018). Understanding Effective Altruism and its challenges. In D. Boonin (ed.), *The Palgrave Handbook of Philosophy and Public Policy* (pp. 441–453). Palgrave Macmillan.

MacAskill, W. (2019). Longtermism. *Effective Altruism Forum*. Available at: https://forum.effectivealtruism.org/posts/qZyshHCNkjs3TvSem/longtermism.

MacAskill, W. (2020). Are we living at the Hinge of history? Global Priorities Institute. Available at: https://globalprioritiesinstitute.org/wp-content/uploads/William-MacAskill_Are-we-living-at-the-hinge-of-history.pdf.

MacAskill, W. (2022a). EA and the current funding situation. *Effective Altruism Forum*. Available at: https://forum.effectivealtruism.org/posts/cfdnJ3sDbCSkShiSZ/ea-and-the-current-funding-situation.

MacAskill, W. (2022b). *What We Owe the Future*. New York: Basic Books.

MacAskill, W., Bykvist, K. and Ord, T. (2020). *Moral Uncertainty*. Oxford: Oxford University Press.

MacKenzie, M. K. (2020). There is no such thing as a short-term issue. *Futures*, 125, 102652. https://doi.org/10.1016/j.futures.2020.102652.

Malde, J. (2021). Possible misconceptions about (strong) longtermism. *Effective Altruism Forum*. Available at: https://forum.effectivealtruism.org/posts/ocmEFL2uDSMzvwL8P/possible-misconceptions-about-strong-longtermism.

Manheim, D. (2020). The fragile world hypothesis: Complexity, fragility, and systemic existential risk. *Futures*, 122, 102570. https://doi.org/10.1016/j.futures.2020.102570.

Manheim, D. (2021a). Towards a weaker longtermism. *Effective Altruism Forum*. Available at: https://forum.effectivealtruism.org/posts/fStCX6RXmgxkTBe73/towards-a-weaker-longtermism.

Manheim, D. (2021b). Noticing the skulls, longtermism edition. *Effective Altruism Forum*. Available at: https://forum.effectivealtruism.org/posts/ZcpZEXEFZ5oLHTnr9/noticing-the-skulls-longtermism-edition.

Masrani, V. (2020). A case against strong longtermism. *Effective Altruism Forum*. Available at: https://forum.effectivealtruism.org/posts/7MPTzAnPtu5HK esMX/a-case-against-strong-longtermism.

McLaughlin, P. (2022). A History of the concept of existential risk. Thesis. Cambridge University.

Mitchell, A. and Chaudhury, A. (2020). Worlding beyond "the" "end" of "the World": White Apocalyptic Visions and BIPOC Futurisms. *Int. Relat.*, 34(3), 309–332. https://doi.org/10.1177/0047117820948936.

Morgan, M. G. (2014). Use (and Abuse) of expert elicitation in support of decision making for public policy. *Proc. Natl. Acad. Sci. USA*, 111(20), 7176–7184. https://doi.org/10.1073/pnas.1319946111.

Moynihan, T. (2020). *X-Risk: How Humanity Discovered Its Own Extinction*. Cambridge, Massachusetts: MIT Press.

Muelhauser, L. (2013). We are now the "Machine Intelligence Research Institute" (MIRI). Machine Intelligence Research Institute. Available at: https://intelligence.org/2013/01/30/we-are-now-the-machine-intelligence-research-institute-miri/.

Ngo, R. (2019). The career and the community. *Effective Altruism Forum*. Available at: https://forum.effectivealtruism.org/posts/Lms9WjQawfqER wjBS/the-career-and-the-community.

Noun, A. and Chyba, C. F. (2008). Biotechnology and biosecurity. In N. Bostrom and M. M. Cirkovic, (eds.), *Global Catastrophic Risks*. Oxford: Oxford University Press.

Ó hÉigeartaigh, S. S., *et al.* (2021). Video: Doom and doubt: Uncertain futures & open questions about existential risk. CSER. https://www.cser.ac.uk/resources/doom-and-doubt/.

Ord, T. (2020). *The Precipice: Existential Risk and the Future of Humanity*. Bloomsbury Publishing.

Ord, T., Mercer, A. and Dannreuther, S. (2021). Future proof: The opportunity to transform the UK's resilience to extreme risks. The Centre for Long-Term Resilience. Available at: https://www.longtermresilience.org/futureproof.

Oreskes, N. and Conway, E. M. (2011). *Merchants of Doubt: How a Handful of Scientists Obscured the Truth on Issues from Tobacco Smoke to Global Warming*. Bloomsbury Publishing.

Owe, A. (2022). Greening the Universe: The case for ecocentric space expansion. In J. S. J. Schwartz, *et al.* (eds.), *Reclaiming Space: Progressive and Multicultural Visions of Space Exploration*. Oxford: Oxford University Press.

Peters, J. D. (2017). You mean my whole fallacy is wrong: On technological determinism. *Repres*, 140(1), 10–26. https://doi.org/10.1525/rep.2017.140.1.10.

Peutherer, N. (2021). Ethics of the far future: Why longtermism does not imply anti-capitalism. *Inquiries J.*, 13(11). http://www.inquiriesjournal.com/articles/1924/ethics-of-the-far-future-why-longtermism-does-not-imply-anti-capitalism.

Pielke, R. A. (2007). *The Honest Broker: Making Sense of Science in Policy and Politics*. Cambridge: Cambridge University Press. https://doi.org/10.1017/CBO9780511818110.

Popper, K. R. (2008). *The Logic of Scientific Discovery*. London: Routledge Classics.

Prunkl, C. and Whittlestone, J. (2020). Beyond near- and long-term: Towards a clearer account of research priorities in AI ethics and society. *Proceedings of AIES 2020 — Proceedings of AAAI/ACM Conference on AI Ethics Society*, pp. 138–143. https://doi.org/10.1145/3375627.3375803.

Rees, M. (2021). Martin Rees and Steven Pinker: Wagering on catastrophe. *New Statesman*. Available at: https://www.newstatesman.com/politics/uk/2021/06/martin-rees-and-steven-pinker-wagering-catastrophe.

Rees, M. and Pinker, S. (2017). A bioterror or bioerror will lead to one million casualties in a single event within a six month period starting no later than December 31 2020. *Long Bets* (blog). Available at: https://longbets.org/9/.

Rhodes, C. (2020). Scientific freedom and responsibility in a biosecurity context. In S. Giordano, J. Harris, and L. Piccirillo (eds.), *The Freedom of Scientific Research* (pp. 105–120). Manchester: Manchester University Press. https://library.oapen.org/bitstream/handle/20.500.12657/44038/external_content.pdf?sequence=1#page=132.

Rios Rojas, C., *et al.* (2021). Foresight for unknown, long-term and emerging risks, approaches and recommendations. Report. Available at: https://doi.org/10.17863/CAM.64582.

Rios Rojas, C, Richards, C., and Rhodes, C. (2021). *Pathways to Linking Science and Policy in the Field of Global Risk*. Centre for the Study of Existential Risk. Report. Available at: https://www.cser.ac.uk/news/new-report-pathways-linking-science-and-policy-fie/.

Roser, M. (2021). The world is much better; The world is awful; The world can be much better. *Our World In Data*. Available at: https://ourworldindata.org/much-better-awful-can-be-better.

Rowell, D. and Connelly, L. B. (2012). A history of the term 'moral hazard'. *J. Risk Insur.*, 79(4), 1051–1075. https://doi.org/10.1111/j.1539-6975.2011.01448.x.

Sagan, C. (1983). Nuclear war and climatic catastrophe: Some policy implications. *Foreign Aff.* 62 (2), 257–292. https://doi.org/10.2307/20041818.

Sandberg, A. (2018). Human extinction from natural hazard events. In *Oxford Research Encyclopaedia of Natural Hazard Science*. https://doi.org/10.1093/acrefore/9780199389407.013.293.

Sandberg, A. (23 May 2021) (Manuscript) Grand futures: Visions and limits of what can be achieved. Working manuscript.

Schmidt, A. T. and Juijn, D. (2021). Economic Inequality and the Long-Term Future. *Global Priorities Institute Working Paper No. 4-2021*. Available at: https://globalprioritiesinstitute.org/economic-inequality-and-the-long-term-future-andreas-t-schmidt-university-of-groningen-and-daan-juijn-ce-delft/.

Schot, J. and Steinmueller, W. E. (2018). Three frames for innovation policy: R&D, systems of innovation and transformative change. *Res. Policy*, 47(9), 1554–1567. https://doi.org/10.1016/j.respol.2018.08.011.

Schubert, S. (2017). Understanding cause-neutrality. Centre for Effective Altruism. Available at: https://www.centreforeffectivealtruism.org/blog/understanding-cause-neutrality.

Schubert, S., Caviola, L., and Faber, N. S. (2019). The psychology of existential risk: Moral judgments about human extinction. *Sci. Rep.*, 9(1), 1–8. https://doi.org/10.1038/s41598-019-50145-9.

Schwitzgebel, E., *et al.* (2021). The racial, ethnic, and gender diversity of philosophy students and faculty in the United States: Recent data from several sources. *Philos. Mag. (Lond.)*, 93, 71–90.

Sears, N. A. (2020). Existential security: Towards a security framework for the survival of humanity. *Glob. Policy*, 11(2), 255–266. https://doi.org/10.1111/1758-5899.12800.

Sempere, N. (2022). Some data on the stock of EATM funding. Measure is Unceasing. https://nunosempere.com/blog/2022/11/20/brief-update-ea-funding/.

Shekhtman, L. M., Gates, A. J., and Barabási, A.-L. (2022). Mapping philanthropic support of science. *arXiv.* https://doi.org/10.48550/arXiv.2206.10661.

Shimi, A. (2020). Will OpenAI's work unintentionally increase existential risks related to AI? *LessWrong.* Available at: https://www.lesswrong.com/posts/CD8gcugDu5z2Eeq7k/will-openai-s-work-unintentionally-increase-existential.

Slate Star Codex. (2020). Reader survey 2020. Available at: https://docs.google.com/forms/d/e/1FAIpQLSd4I-x9oArWW1Tz5mEK4uHmxcJzVKGA28RfKPsDvW8hzZNViw/viewanalytics.

Sohn, E. (2023). How philanthropy can nurture your research. *Nature.* https://doi.org/10.1038/d41586-023-00077-2.

Stix, C. and Maas, M. M. (2021). Bridging the gap: The case for an 'incompletely theorized agreement' on AI policy. *AI Ethics*, 1, 261–271. https://doi.org/10.1007/s43681-020-00037-w.

Sundaram, L., Maas, M. M., and Beard, S. J. (2022). Seven questions for existential risk studies. *SSRN.* https://doi.org/10.2139/ssrn.4118618.

Tarsley, C. (2019). The epistemic challenge to longtermism. *GPI Working Paper No. 10-2019*. Global Priorities Institute. Available at: https://globalprioritiesinstitute.org/christian-tarsney-the-epistemic-challenge-to-long termism/.

Tetlock, P. E. and Gardner, D. (2016). *Superforecasting: The Art and Science of Prediction*. USA: Broadway Books.

Tetlock, P. E., Mellers, B. A. and Scoblic, J. P. (2017). Bringing probability judgments into policy debates via forecasting tournaments. *Science*, 355(6324), 481–483.

Todd, B. (2017). The case for reducing existential risks. 80,000 hours. Available at: https://80000hours.org/articles/existential-risks/.

Todd, B. (2021). How are resources in EA allocated across issues? *Effective Altruism Forum*. Available at: https://forum.effectivealtruism.org/posts/nws5pai9AB6dCQqxq/how-are-resources-in-ea-allocated-across-issues.

United Nations. (2021). Our common agenda: Report of the secretary-general. United Nations. Available at: https://www.un.org/en/content/common-agenda-report/.

University of Cambridge. (2020). 2018–2019 equality and diversity information report. Available at: https://www.equality.admin.cam.ac.uk/files/equality_information_report_18-19_.pdf.

Vaintrob, L. (2021). Humanities research ideas for longtermists. *Effective Altruism Forum*. Available at: https://forum.effectivealtruism.org/posts/oTJ5vMNwdWiHj2iKL/humanities-research-ideas-for-longtermists.

Vaintrob, L. (2022). Against "longtermist" as an identity. *Effective Altruism Forum*. Available at: https://forum.effectivealtruism.org/posts/FkFTXKeFxwcGiBTwk/against-longtermist-as-an-identity.

Wagner, G. and Merk, C. (2019). Moral hazard and solar geoengineering. In *Governance of the Deployment of Solar Geoengineering*. Harvard Project on Climate Agreements, pp. 135–139. Available at: https://gwagner.com/wp-content/uploads/Wagner-Merk-2019-Moral-Hazard-and-Solar-Geoengineering-brief.pdf.

Walker, P. F. and Koblentz, G. D. (2017). Can Bill Gates rescue the bioweapons convention? *Bulletin of the Atomic Scientists*. Available at: https://thebulletin.org/2017/04/can-bill-gates-rescue-the-bioweapons-convention/.

Walsh, J. (2021). Elon Musk's false Covid predictions: A timeline. *Forbes*. Available at: https://www.forbes.com/sites/joewalsh/2021/03/13/elon-musks-false-covid-predictions-a-timeline/?sh=5b2f8f965b6d.

Weiss, C. H. (1979). The many meanings of research utilization. *Public Admin. Rev.*, 39(5), 426–431. https://doi.org/10.2307/3109916.

Wells-Jensen, S. and Beard, S. J. (2023). We have to include everyone: Enabling humanity to reduce existential risk. In S. J. Beard *et al.* (eds.), *The Era of Global Risk: An introduction of Existential Risk Studies* (pp. 101–122). Cambridge: Open Book Publishers.

Wiblin, R. (2016). A framework for comparing global problems in terms of expected impact. 80,000 Hours. Available at: https://80000hours.org/articles/problem-framework/.

Wiblin, R. and Harris, K. (2021). Carl Shulman on the common-sense case for existential risk work and its practical implications. *80,000 Hours Podcast*. Available at: https://80000hours.org/podcast/episodes/carl-shulman-common-sense-case-existential-risks/.

234 L. S. Sundaram et al.

Williams, R. and Yampolskiy, R. (2021). Understanding and avoiding AI failures: A practical guide. *Philosophies*, 6(3), 53–78. https://doi.org/10.3390/philosophies6030053.

Winter, C., *et al.* (2021). Legal priorities research: A research agenda. Legal Priorities Project. Available at: https://www.legalpriorities.org/research_agenda.pdf.

Wintle B. C., *et al.* (2017). A transatlantic perspective on 20 emerging issues in biological engineering. *elife*. 6, e30247.

Yudkowsky, E. (2007). Making Beliefs Pay Rent (in Anticipated Experiences). *LessWrong*. Available at: https://www.lesswrong.com/posts/a7n8GdKiAZRX86T5A/making-beliefs-pay-rent-in-anticipated-experiences.

Yudkowsky, E. (2017). Security mindset and ordinary Paranoia. Machine Intelligence Research Institute. Available at: https://intelligence.org/2017/11/25/security-mindset-ordinary-paranoia/.

Index